Mind Maps in Biochemistry

Authored by

Simmi Kharb

Department of Biochemistry, and Nodal Officer, Multidisciplinary Research Unit, Pt. B. D. Sharma Postgraduate Institute of Medical Sciences (PGIMS), Rohtak, Haryana 124001, India

Mind Maps in Biochemistry

Editor: Simmi Kharb

ISBN (Online): 978-981-14-7790-4

ISBN (Print): 978-981-14-7788-1

ISBN (Paperback): 978-981-14-7789-8

Published by Bentham Science Publishers Pte. Ltd. Singapore.

Bentham Science Publishers Pte. Ltd.
80 Robinson Road #02-00
Singapore 068898
Singapore
Email: subscriptions@benthamscience.net

CONTENTS

FOREWORD

Undersigned is excited to record personal impressions on book, "Mindmaps in Biochemistry" authored by Dr. Simmi Kharb working at Department of Biochemistry, Pt. B. D. Sharma Postgraduate Institute of Medical Sciences (PGIMS), Rohtak, Haryana, India as a Professor and Nodal Officer at the Multidisciplinary Research Unit of the Institute.

The first textbook on Biochemistry was written by Alexander Thomas Cameron in 1928. Since then, ginormous development has taken place in the field of Biochemistry into different scientific canals. On one hand the whole genome has been unfurled and on other hand, there has been great demand of traditional knowledge. When the world is passing through turmoil of knowledge bank, there has been great development in understanding the 'Chemistry of Biological Systems'. Professor Simmi Kharb has generated literature by virtue of her book in the arena of Biochemistry with an aim (ambition in mind) that the present day students studying Biochemistry need comprehensive information at one place in understandable language that can act as unstoppable orientation in young minds that would lead to formation of strong grass root for becoming passionate Biochemists.

It was being felt by the teachers and students of Biochemistry that a composite collection of Biochemical Principles which can find place in the minds of learners just like map, was missing in the literate world. Dr. Simmi Kharb accepted the challenge of producing a book on Biochemistry that has simple, consolidated, easy to read and retain, all in one, mixture of basic and advanced knowledge module.

The issues covered in this book are: Biochemical concepts, Organization of chemical and biological approach, Principles in balance of biological systems, Acquiring meaningful understanding of Biochemistry, Real-time world relevance and Problem-solving mechanisms. The book runs through cell to body.

I am confident that this book shall make its place in libraries, minds of teachers and vocal cord of students all over the world.

Gajendra Singh, Ph.D
Professor and Dean
Faculty of Pharmaceutical Sciences
Pt. BD Sharma University of Health Sciences
Rohtak 124001
Haryana
India

PREFACE

Students often view Biochemistry as a pile of facts or equations to be memorized rather than as concepts to be understood. The author proposes to create a series of concept and knowledge maps about the biochemical contents to illustrate graphically the relationships between the ideas presented in a given proposed chapter, as well as to show how information can be grouped or organized.

In order to facilitate the student's understanding of the metabolic pathways and providing greater interaction with the contents, an approach to better learning metabolic diagrams will be developed through:

i. Flow diagrams and illustrations: showing substrates and enzymes of the metabolic pathways, their control, inhibition, role of vitamins in their correct functioning, and their connection with other systems
ii. Reading the functions and characteristics of metabolic pathways in illustrations
iii. Solving of essay and multiple-choice questions
iv. Recent advances and applied aspects of biochemistry, applied therapeutics and microbiology will be discussed

Also, students will be encouraged to make their own flow diagrams and tables and sample questions will be provided to improve their analytical skills.
As a learning tool, it can be used by the student to, *e.g.*, making notes, solving problems, planning the study and/or writing of essays, preparing for examinations, and identifying the connection of topics.

ACKNOWLEDGMENT & CONFLICT OF INTEREST

No potential conflict of interest is declared by the author. It is also declared the complete work is an individual effort by the author and there was no financial/ administrative/ academic support availed from any individual/ institution /organization.

Simmi Kharb
Department of Biochemistry
Nodal Officer, Multidisciplinary Research Unit
Pt. B. D. Sharma Postgraduate Institute of Medical Sciences (PGIMS)
Rohtak, Haryana 124001
India

CHAPTER 1

Cell, Plasma Membrane, Membrane Transport

LEARNING OBJECTIVES:

- Illustrate the cellular components and their significance.
- Appraise the structure and function of the cell membrane.
- Explain the transport mechanisms operating across the plasma membrane.

Keywords:

Active transport, Eukaryotic cells, Prokaryotic cell, Plasma membrane, Passive transport, Sub-cellular component.

KEY FEATURES OF CELL

1. Cell: structural and functional unit of life.
2. Most chemical reactions take place within cells.
3. Cells are of two types:
   i. Prokaryotic cell
   ii. Eukaryotic cell

- The human body is composed of 10^{14} cells.

Cells possess a genetic program and the mean to use it: capable of producing more of themselves.

Cells acquire and utilize energy, carry out a variety of chemical reactions: metabolism.

Cells engage in mechanical activity able to respond to stimulant.

PROKARYOTIC CELL

Lack well-defined nucleus.

Contain rigid cell wall.

EUKARYOTIC CELL

Cells are highly complex and organized.

Contain well defined nucleus, more complicated internal structure.

Comparison	**Prokaryotic Cell**	**Eukaryotic**
Size	Small 1-10μ	Large 10-100μ

Simmi Kharb

Cell Wall	Present	As extracellular matrix
Cell Membrane	Rigid, cell wall	Flexible, plasma membrane
Nucleus	No well-defined nucleus	Well-defined nucleus, contain membrane
DNA	Found as nucleoid	DNA associated with histones
Histones	Absent	Present
Nucleolus	Absent	Present
Genome	Single, circular chromosome	Multiple chromosomes
Metabolism	Both aerobic and anaerobic	Aerobic
Respiratory Enzymes	Located in the plasma membrane	Located in mitochondria
Cell Division	Cleavage, fission	Mitosis, meiosis
Cytoplasm	Lack organelle and cytoskeleton	Contain organelle and cytoskeleton

CYTOSOL

Fluid compartment: soluble part, viscous gel.

Represents 50-60% of total cell volume

It is in contact with all sub-cellular organelle.

Contains: enzyme, metabolites and salts.

Functions

Carbohydrate metabolism

Fat metabolism

Nucleotide metabolism

COMPOSITION OF CYTOSOL

1. **Proteins:**
   a. Enzymes
   i. Carbohydrate metabolism
   ii. Protein metabolism
   iii. Fat metabolism
   b. Transporter, carrier proteins
2. **Metabolites:**
   Of carbohydrate and amino acid metabolism.
3. **Other Molecules:**
   Na^+, K^+, Ca^{2+}, Mg^{2+}, HCO_3^-, Cl^-, PO_4^-

METABOLIC PATHWAYS IN CYTOSOL

1. Carbohydrate Metabolism

a. Glycolysis

b. PPP

c. Glycogen

2. Fatty Acid Synthesis (*de novo*)

3. Amino Acid Metabolism

a. Oxidation

b. Deamination

c. Decarboxylation

d. Transamination

4. Pathway present:

a. Initial step of:

i. Heme synthesis

ii. Urea synthesis

b. ***Nucleotide synthesis:*** many steps occur in cytosol:

i. CPS II

ii. ATCase

CYTOSKELETON

Composed of:

- ***Actin filament:*** 7 nm thick, attached to adherens junction, reacts with myosin filament during contraction.
- ***Intermediate filament:*** 10 nm thick, attaches to desmosomes and hemidesmosomes.
- ***Microtubules*:** 25 nm thick, hollow, composed *of α and β tubulin* subunits.

Functions

- Allows cells to move and adopt different shapes.
- Play role in cell division: microtubules form mitotic spindle.
- Cell movement, determining cell shape, axonal transport.

Cilia

Move fluid over epithelial surfaces *e.g.* respiratory tract

Applied aspect

Immotile cilia can cause Kartagener's syndrome:

Sinusitis, Bronchitis, Situs inverses

Immotile cilia caused by:

Fault in motor protein *dynein:* responsible for the rhythmic movement of cilia and flagella.

CELL JUNCTIONS

Four classes of junction that form epithelial cells:

tight, adherens, desmosomes, gap junctions

NUCLEUS

Large fragmental structure bounded by a double membrane, contains nucleolus.

Generally, cells in human body have a single nucleus except**:**

Skeletal muscle cells have many nuclei.

Mature RBC have no nucleus.

Flagella of spermatozoa Used as means of propulsion	**Function** Nucleus is the primary store of DNA and mitochondria contain their own DNA.
GOLGI APPARATUS Flattened, disc like membrane bound organelle. Responsible for modification of lipid and proteins. Sorts, modifies and transports them to targets. Contain many distinct cisternae arranged in stacks. Each stack has two faces: Trans: aligned towards plasma membrane, serve as entry point. Cis: adjacent to rough endoplasmic reticulum, serves as exit for modified proteins, functions as a sorting station. Release secretary proteins as vesicles. Specialized secretary cells contain large amounts of Golgi apparatus. **Secretion of Vesicles** They pass through plasma membrane to surrounding media Examples: Ig from secretary vesicles: (products stored and released under appropriate stimuli) Trypsinogen (pancreatic acinar cells) Insulin (β cells Langerhans). Targeted to lysosomes after tagging it to mannose of N-oligosaccharide. Reach plasma membrane, form integral part of it.	**ENDOPLASMIC RETICULUM (ER)** System of membranes that enclose a lumen forming a network of flattened sacs (cisternae). Two types: o Smooth (SER) o Rough (RER) Composed of tubular network throughout cytoplasm. Involved in modification and detoxification and hydrophobic compound. *E.g.*, Hepatocytes have large number of SER. Predominates in cells specialization in lipid metabolism. **Functions** Synthesis of: • Lipoprotein • TAG • Phospholipids • Part of cholesterol • Bile salts • Steroid hormone synthesis Detoxification: of organic compounds: barbiturate, ethanol Conjugation reactions

RER (Translational ER)	**MICROSOMES**
RER is continuous with outer membrane of nuclear envelope. Has grainy appearance on electron microscopy (high concentration of membrane bound ribosomes). Involved in synthesis of proteins destined to be targeted to organelles or to function outside the cell. Disulphide bond formed and glycosylation occurs in RER. Cells of pancreas salivary gland have extensive RER. ***Functions of RER*** Synthesis of protein: mucoprotein, polypeptide hormone, antibodies, plasma protein (albumin) (insulin, glucagon). Synthesis of integral membrane proteins other than those of mitochondria.	When cells are fractionated complex ER network is not isolated as a whole. It is disputed in many places. These membranes automatically reassemble to form *Microsomes*.

MITOCHONDRIA	**Maternal Inheritance of mt DNA**
Cylindrical, vary in size 0.5-2μm, appear as sausage-shaped bodies. Bound by a double membrane, dividing interior into 2 spaces: inter membranous space, matrix. Inner membrane has many projections forming folds called cristae that increase surface area	Mt DNA and ribosomes are located in matrix: equipped with independent protein synthesizing machinery. Supplies up to 35% of protein of mitochondria rest come from nucleus.

<table>
<tr><td rowspan="2">Contain their own DNA of maternal origin.

Mitochondrial proteins derived from nuclear DNA.

Neutral pH in matrix

Biochemistry of mitochondria

Functions

Power Plant of cell: principal producer of ATP, NADH

Cycle: Operation of citric acid cycle and β-oxidation of FA

Storage of calcium ions.</td><td>Mitochondria have evolved from aerobic bacteria by endosymbiosis.</td></tr>
<tr><td>Distribution

Liver or stomach: distributed haphazardly

Muscle – packed systematically around myofibrils

Adipocytes – mitochondria surround large globules of TAG.</td></tr>
</table>

Enzymes/Proteins and Constituents in Mitochondria Compartment

OMM	**Intermembraneous**	**IMM Outer Inner**	**Matrix**
MAO	Adenylate Ainase	GA-3P DH SDH	TCA β-oxidation
Acetyl CoA synthetase	Creatine kinase	Energy of respiratory chain	Glutamate deH
PLA_2	Nucleotide metabolism	β-OH butyrate deH	Cis aconitase Replication of mt DNA Protein Synthesis Transport system Pyruvate Malate

NUCLEUS	**Clinical Consequence**
Largest cellular organelle. All cells in body contain nucleus except RBCs. Bound by a double (each 8 nm thick) membrane known as nuclear envelope. Separated by 20-40 nm perinuclear space. Outer nuclear membrane (ONM) continuous with membrane of ER: dotted with ribosomes ONM and inner nuclear membrane (INM) connected by nuclear pairs arranged in hexagonal array of diameter: 60-90 nm. On inner side of INM lies nuclear lamina which is composed of fibrous network. In some cells, nucleus occupies most of available space, *e.g.* small lymphocyte, spermatozoa. Nucleus contains DNA**:** repository of genetic information. DNA associated with histones forms: NUCLEOSOMES. A single chromosome is composed of about a million nucleosomes. Nucleus possesses entire glycolytic sequence of enzymes. Some nuclei *e.g.* lymphocytes can carry out oxidative reactions of citric acid cycle. Chromatin: assembly of nucleosomes, composed of DNA, histones, Non-histone proteins, RNA.	Regulation of DNA expression and function is severally disturbed in serious pathological conditions. *E.g.* cancer leading on to uncontrolled production of a particular hormone from normal gland or ectopic production. **Nucleolus** Sub-cellular organelle located in nucleus. Bead like structure. Contain RNA particularly, directly rRNA reflects synthetic activity of cell. Disappears during cell division at prophase becomes associated with their satellite chromosomes. 15 nm in diameter. Site of synthesis of most rRNA and where ribosome assembly begins. *Enzymes present:* • RNA polymerase • Ribonuclease. **NUCLEOPLASM** Ground substance of nucleus, contains:

Humans have *46 chromosomes* compactly packed in the nucleus. *Human haploid genome has 2.8 x 10^9 bp.*	□ Enzymes of DNA and RNA synthesis. □ Co factors and coenzymes like ATP and acetyl CoA. □ Nucleic acids, proteins, lipids and minerals.

LYSOSOMES Membrane bound sacs that contain hydrolytic enzymes pH of lysosomal matrix: 5.0. Most abundant in macrophages fewer in lymphocytes. Lysosomes receive cellular and endocytosed proteins and lipids that need digesting and the resulting metabolites are transported either by vesicles or directly across the membrane. Control intracellular digestion of macromolecules. All acid hydrolases in lysosomes are maintained at a pH 5.0 and are inactive in neutral pH of cytosol. Suicide bag of cell. **SECONDARY LYSOSOME** Large irregularly- shaped vesicles in which digestion of material is under way. Appear as a result of fusion of primary lysosomes with other membrane organelles. Contain particles or membrane in the process of being digested. Endocytic vesicles and phagosomes are fused with lysosomes to form secondary lysosomes.	**Lysosomal Enzymes** Protein in nature. Synthesized in RER. Receive n-linked oligosaccharide. Transported to SER by pinching out as vesicle, transported to Golgi complex, where mannose residue get phosphorylated and then directed to lysosomes. **Lysosomal Enzymes: Types** 1. ***Hydrolases:*** Polysaccharide hydrolyzing enzyme a. α-glucosidase b. β-galactosidase c. α-mannosidase d. Hyaluronidase e. Lysozyme (bacterial cell wall) f. Fucosidase g. Sialidase 2. ***Protease:*** Protein hydrolyzing enzymes a. Cathepsin

LIFE CYCLE OF LYSOSOMES Synthesized in Golgi apparatus as vesicles spherical. Do not contain particulate or membrane debris. Here enzymes are not involved in digestive process. **Two Pathways of Life Cycle of Lysosomes** **Heterophagy** Endocytosis of ingested particle form an intracellular vesicle which fuses with a primary lysosome to form secondary lysosome. Here, active digestion of ingested particle by lysosomal enzyme occurs, amino acids or monosaccharide are released into cytoplasm for further metabolism. Undigested material so remains forms tertiary lysosomes, which may be expelled by exocytosis. **Autophagy** Similar to heterophagy, but the particle trapped in vacuole is usually another sub-cellular particle or organelle. If a cell is deprived of nutrients, a marked increase in autophagy occurs.	b. Collagenase c. Peptidase ***3. Nuclease*** a. Acid ribonuclease b. Acid deoxyribonuclease ***4. Phosphatases*** a. Acid phosphatase b. Acid phosphodiesterase ***5. Lipid hydrolyzing enzymes*** a. Acid lipase b. Lipase c. Phospholipase d. Esterase ***6. Sulfatase*** a. Iduronate sulfatase b. Heparan N-sulfatase c. β-galactosidase
Examples of Heterophagy 1. In proximal tubule: albumin and Hb degraded by lysosomal cathepsins. 2. Lymphoid tissue: spleen and thymus: Endocytosis of: • Foreign material entering blood stream by macrophages • Damaged or aged RBCs in spleen	**Clinical Applications** 1. **Gout:** Phagocytosed urate crystals damage vacuole and release hydrolytic enzymes causing inflammation. 2. **Tumour Metastasis** Certain cancer cells liberate cathepsins (which are normally

- Damaged cells of spleen and thymus.
- Endocytosis of antigen inside macrophage lysosomes before production of antibodies

3. **Nervous Tissue:** Autophagic role by removing damaged proteins in neurons. Heterophagic role under pathological conditions

4. **Bone:** Lysosomes play an important role in *resorbing* and bone, *remodeling* the cells of bone by osteoclasts enzymes are hyaluronidase, peptidase, and collagenase

5. **Uterus:**

    a. Autophagy activity increases at the end of cycle.

    b. Post-partum involution of uterus occurs by both autophagy and heterophagy.

6. **Mammary Gland:** Involution of mammary gland following cessation of lactation, autophagy reduce cytoplasmic content and heterophagy removes dead cell and milk.

restricted to interior of lysosomes) which hydrolyze collagen and elastin, degrade basal lamina and tumor cells move out causing distant metastases.

3. **Lysosomal Storage Disease**

    a. Tay–Sachs GM2-Gangliosidosis: Autosomal recessive, neurological problem and early death.

    b. I-cell disease (inclusion cell disease):

        - Enzymes which phosphorylate mannose residue of glycoprotein in Golgi complex to target it to lysosome are ABSENT.

        - Severe neurological and bone deformity.

4. **Lysosomal Acid Lipase Deficiency**

    Autosomal recessive. Deposition of TAG and Cholesterol esters in tissues particularly liver. Early onset of atherosclerosis.

5. **Apoptosis**

    Lysosomes have as role in apoptosis

6. **Glycogen Storage Disease Type-II**

    Acid maltase deficiency. Glycogen particles taken up by autophagy and

	cannot be digested by autophagy and accumulate in lysosomes. 7. **Hurler's Disease** Deposition of GAG: dermatan sulfate and L-iduronidase deficient in lysosomes.
PEROXISOMES All animal cells except RBCs contain peroxisomes. Single membrane cellular organelles having fine granular matrix. Also known as microbodies, 0.3-1.5 mm in diameter. Prominent in WBCs and platelets, contain oxidative enzymes involving H_2O_2. Contain enzymes that oxidize D-amino acid and uric acid, Long chain FA oxidation, synthesis of glycerol lipids and plasmalogens and H_2O_2 produced in these reactions are detoxified by catalase.	**Clinical Applications** ***Zellweger Syndrome*** • Autosomal recessive. Absence of functional peroxisome. Very long chain FA accumulate • Bile acid synthesis abnormality. Dying in one year. ***Other Defects*** Neonatal

METABOLIC FUNCTIONS OF SUBCELLULAR ORGANELLES

Organelle	Functions	
Nucleus	DNA replication	Transcription
ER	Biosynthesis of proteins, glycoproteins, lipoproteins.	
	Ethanol oxidation	FA desaturation
	Drug metabolism	Synthesis of cholesterol
Golgi apparatus	Maturation of synthesized proteins	

Lysosome	Degradation of proteins, carbohydrates, lipids, nucleotides	
Mitochondria	ET chain ATP generation TCA cycle	β oxidation FA KB production Heme, urea synthesis (part)
Cytosol	Protein synthesis Glycolysis Glycogen metabolism HMP shunt	Transamination FA synthesis Cholesterol synthesis (part) Purine, pyrimidine synthesis (part)

HOW TO STUDY CELLS

Cell Fractionation	**Sub-Cellular Fractionation (SCF)**
• To separate major organelles of the cells so that their individual functions can be studied. • Uses ultracentrifuge. • Fractionation begins with homogenization, gently disrupting the cell. • Mixture is spun in centrifuge to separate heavier pieces into the pellet. • At higher speed and longer durations, smaller organelles can be collected in subsequent pellets.	Breaking open of a cell (by homogenization) and separation of organelles from one another by centrifugation. ***Steps*** o Extraction o Homogenization o Centrifugation o Then subject it to: □ EM □ Marker enzyme ***Application: SCF*** Major component of experimental approach and functions of organelles have been elucidated.

Use	
• Cell fractionation prepares quantities of specific cell components. • Functions, and structure of organelle can be studied.	
Whole Cells Can also be used to Study Biochemical Processes 1. Purification of cells: on basis of density. 2. Flow cytometry: separates different cell types. 3. Fluorescence-activated cell sorter (FACS): can select one cell from thousands of other cells. *E.g.* WBC.	

MARKER ENZYMES

A marker is a chemical or enzyme that is almost exclusively combined to one particular organelle.

Organelle	Marker
Nucleus	DNA FMN Adenyl transferase
Mitochondria	MAO
OMM	Sulfite oxidase
Intermembraneous	Cyt C oxidase, SDH
IMM	Glutamate deH
Matrix	High content of RNA
Ribosome	Cathepsin, acid phosphatase
Lysosomes	Galactosyl transferase

Golgi apparatus	Catalase, uric acid oxidase
Peroxisome	D-amino acid oxidase
Cytosol Microsome (ER)	LDH, G-6-Pase, NADPH –cyt reductase

PLASMA MEMBRANE, TRANSPORT ACROSS MEMBRANE

Fact File: Plasma Membrane

- Defines the boundary of cell and forms boundaries of organelles.
- Regulates movements of material in and out of the cell.
- Facilitates electrical signalling between cells.

Membrane form Hydrophobic Barriers Around Cells to Control Internal Environment by Restricting Entry and Exit of Molecules:

- It is permeable of water, gases (O_2, CO_2, N_2O) and uncharged molecules (urea, ethanol).
- It is impermeable to uncharged polar molecule (glucose) ions (Na^+, K^+, Cl^-, Ca^{2+}), charged polar molecules (ATP, amino acids, G-6-P).

Movement of Uncharged Molecules and Ions Occurs Freely Through Membrane:

- ***Oxygen molecules being uncharged and dissolve readily in water:*** They can also dissolve in hydrophobic interior of plasma membrane and pass into cytosol through simple diffusion.
- ***CO_2, NO and H_2O molecules and uncharged hormones of steroid family***: They also pass across plasma membrane by simple diffusion.
- ***Charged Ions:*** They cannot dissolve in hydrophobic regions and cannot cross membrane by simple diffusion.

Membrane Functions

1. ***Compartmentalization***

   a. It allows specialized activities to proceed without external interference.

b. It allows independent regulation of cellular activities from one another.

2. ***Forms Framework for Biological Activities***

Provides a framework where various components can have effective interactions.

3. ***Provide Selective Permeability Barrier***

It prevents unrestricted exchange of molecules from one site to another site.

4. ***Transport of Solutes***

It facilitates transport of solutes from one side of membrane to another *e.g.* sugar amino acids, ions.

5. ***Signal Transduction***

Membrane possess receptors to which various ligands can bind to generate a signal that stimulates or inhibits internal activities.

6. ***Cell to cell interaction***

7. ***Energy transduction***

Biological Membranes: Properties:

- Highly viscous and plastic structures.
- Form boundaries around cell and sub-cellular compartments.
- Acts as selective permeability barrier.
- Involved in signalling processes.
- Contain varying amount of lipid and protein and carbohydrates.
- Thickness 60-100 Å .
- Dynamic structures.
- Thermodynamically stable, metabolically active.
- Show chemical asymmetry: two faces of biological membrane differ from one another.

Chemical Composition Biological membrane is composed of phospholipids bilayer and membrane proteins. Proteins in membranes may be tightly bound (integral protein) or peripheral proteins.	
Major components	Lipid Protein
Lipids	Glycerophospholipids (GPL) (Fig. **1.1**) Sphingolipids (SPL) Cholesterol
GPL	Have glycerol back bone to which FA and phosphorylated head group are attached. ***Include:*** Phosphatide, phosphatylcholine, phosphatidylethanolamine, phosphatidylglycerol, phosphatidylinositol, phosphatidylserine, and Diphosphatidyl glycerol. ***Location:*** present exclusively in IMM GPL contain two fatty acyl groups esterified to C-1 and C-2 of glycerol: ***C1***– usually saturated FA is found *e.g.* palmitic, stearic acid and form a straight chain. ***C2***- unsaturated FA is present *e.g.* oleic, linoleic, linolenic acid. Usually cis and produces a kink. ***Plasmalogen:***

	• Ethanolamine, ether/choline esterified to phosphate • Abundant in nervous tissue and heart. Every tissue and cellular membrane have a distinct composition of GPL and a definite pattern of FA composition.

SPL	In Sphingomyelin:
	
	Terminal -OH group of sphingosines is esterified to phosphoryl choline, so that its polar head is similar to PC.
	
	Features
	
	• Contain sphingosine back bone
	• Sphingosine:
	    • Amino alcohol
	    • Linked to FA by amide bond forming ceramide
	• Sphingomyelin has PC esterified to 1-OH group
	• Most abundant sphingolipid in mammalian tissues
	• Lack phosphate, have sugar or primary group of sphingosines

Plasma Membrane: Lipid Bilayer Each layer of plasma membrane lipid bilayer formed by phospholipids arranged with hydrophobic head facing aqueous medium and hydrophobic membrane core is formed by fatty acyl tails.	***Lipid Composition*** The composition of lipids in membrane is variable according to cell type: ***a.*** ***In Plasma Membrane***: Highest concentration of neutral lipids and SPL ***b.*** ***Myelin Membrane:*** rich in SPL and GPL

<table>
<tr><td>Hydrophobic tails have saturated FA and unsaturated FA (forms kinks making membrane more fluid).</td><td rowspan="2">c. <u>Intracellular Membrane</u> primarily contains GPL

d. <u>In Mitochondria, Nucleus and RER:</u> lipid composition is similar

e. <u>IMM:</u> Cardiolipins are present exclusively in IMM and phosphatidylcholine

i. SPL predominant and ethanolamine

ii. GPL are second.

f. <u>In Mitochondria:</u> there is no sphingosine

g. <u>Golgi Apparatus:</u> has high concentration of neutral lipids especially cholesterol</td></tr>
<tr><td>Principal Phospholipids in the Membrane are:

• <u>Glycolipids:</u> Phosphatidylcholine (PC), phosphatidyl ethanolamine (PE) and phosphatidyl serine (PS).

• <u>Sphingolipid:</u> sphingomyelin.</td></tr>
</table>

<table>
<tr><td>Lipid Composition is Asymmetric

• High PC and sphingomyelin content on outer leaflet

• Higher content of PE and PS in the inner leaflet.</td><td rowspan="2">Amphipathic Nature of Membrane Lipids

Membrane lipids contain hydrophobic and hydrophilic regions termed: Amphipathic

Polar head of phospholipids and –OH group of cholesterol are:

• <u>Hydrophilic</u>: favor contact with water

• <u>Hydrophobic Tail:</u> Saturated fatty acids straight chain

<u>Role of Kinking:</u>

• Unsaturated FA make kinked tails

• When more kinks are inserted in tails membranes become more tightly packed and becomes more fluid.</td></tr>
<tr><td>Cholesterol in Membrane:

• Most common sterol in membrane residing mainly in plasma membrane and in mitochondrial, nuclear and Golgi apparatus membrane.

• <u>Amphipathic:</u>

o OH group of cholesterol at interface and remainder within the leaflet.</td></tr>
</table>

o Ring system gives rigidity to membranes.	

Carbohydrates in Membrane

Composition

Oligosaccharides covalently attached to:

- Proteins to form glycoproteins
- Lipids in lesser amount to form glycolipids

Sugars in Glycoprotein and Glycolipid

Glucose, galactose, mannose, fructose, n-acetyl glucosamine (NAc Gln), n-acetyl galactosamine (NAc Gal), Sialic acid.

Location

Carbohydrates are located on:

- External side of plasma membrane
- Luminal side of ER.

Role

- Cell- cell recognition, adhesion, receptor action
- Carbohydrates of glycolipids of RBC plasma membrane determine whether person's blood type A, B, AB or O.

Proteins in Lipid Bilayer

Proteins in lipid bilayer can function as channels or transporters; structural proteins or neurotransmitters.

Membrane form hydrophobic barriers around cells to control internal environment by restricting entry and exit of molecule. Classified as

<table>
<tr><th colspan="2">Integral (intrinsic) Membrane Proteins</th><th>Peripheral (extrinsic) Membrane Proteins</th></tr>
<tr><td colspan="2">Span the lipid bilayer, may span bilayer many time.

Present in membrane.

Amphipathic and globular e.g. transporter, receptors, G protein.

Distributed asymmetrically across bilayer.

Require detergents for their release.</td><td>Can be found on cytosolic or extracellular side of membrane.

Bound to membrane by electrostatic and H- bond interactions with head groups of lipids and integral proteins.

May be anchored to bilayer.

Do not interact with PL of bilayer.

Do not require detergents for their release.</td></tr>
<tr><td colspan="3"><u>Interaction of Proteins with Membrane:</u> Form</td></tr>
<tr><td><u>Membrane- spanning:</u>

α-helices: most common</td><td><u>Channel forming:</u>

β strands: porin protein</td><td><u>Bound to other Membrane Protein:</u>

Ankyrin, PGH2 synthase</td></tr>
</table>

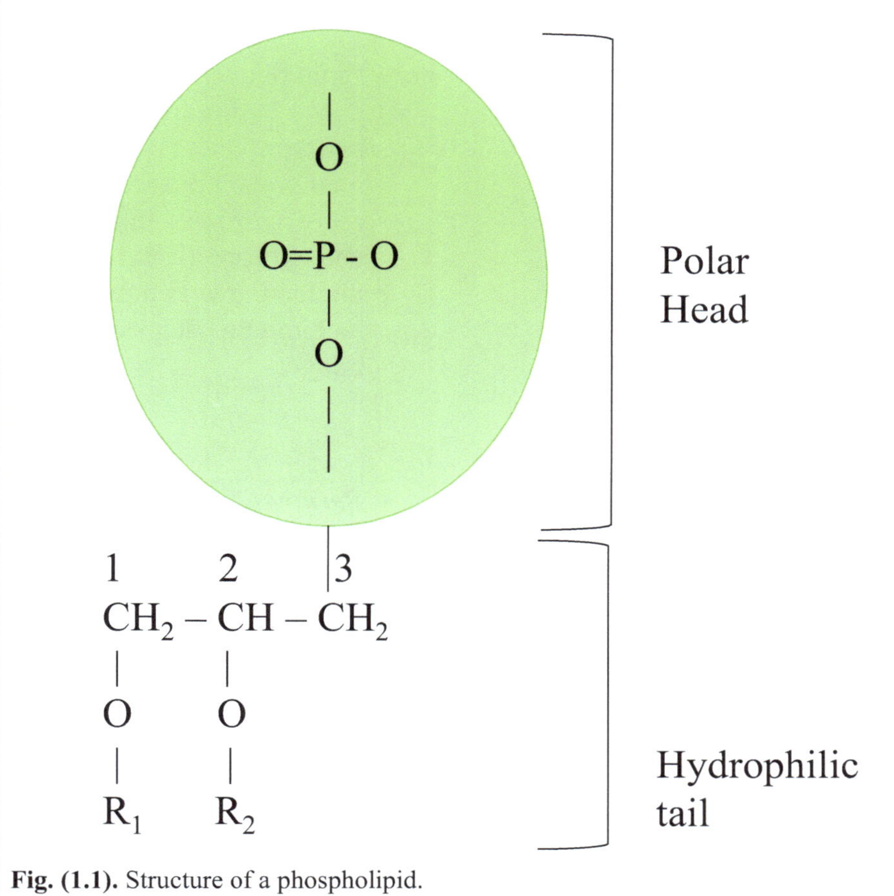

Fig. (1.1). Structure of a phospholipid.

Ankyrin:

Peripheral protein is bound to spectrin (a cytosolic structure) to maintain biconcave shape of RBC

PGH_2 synthase:

Lies along outer surface of membrane

Bound by a set of α-helices

Integral protein

Its localization is crucial for its function:

Arachidonic acid, a substrate generated by hydrolysis of membrane lipids and reaches enzyme through a hydrophobic channel:

This channel is blocked by Aspirin, ibuprofen, thus inhibiting PG synthesis.

Membrane Formation is Consequence of Amphipathic Nature of Molecules

PL and GL (glycolipid) readily form bimolecular sheets in aqueous media.

Molecules with above mentioned preferences can thus arrange themselves in aqueous solution in two ways:

<table>
<tr><td>Micelle

A globular structure of 200 nm size with polar head surrounded by water and carbohydrate tails sequestered inside.</td><td>Lipid Bilayer

Lipid bilayer in formed due to hydrophilic and hydrophobic moieties of membrane lipids resulting in hydrophobic interior and hydrophilic exterior interacting with aqueous medium on each side of the bilayer.</td></tr>
<tr><td>Formation of Lipid Bilayer is

• Self- Assembly Process
• Rapid and Spontaneous in Water
• Dividing Force is Hydrophobic Interactions.</td><td>Permeability

• Steroids traverse bilayer more readily than electrolytes
• Non- lipid soluble molecules can't pass through bilayer and require pores and channels for them</td></tr>
<tr><td>Liposomes

• They are lipid vesicles with aqueous compartments enclosed by a lipid bilayer.
• Important experimental and clinical tool.
• Produced by suspending phospholipids (PC) in aqueous medium and sonicating to give closed vesicles of 500 Å.</td><td>Uses

o To study effect of varying lipid composition on certain membrane functions.
o Various drugs, antibodies or isolated gene/ DNA can be trapped</td></tr>
</table>

	inside them to target them to specific tissues.

BIOLOGICAL MEMBRANE STRUCTURE: FLUID MOSAIC MODEL (Fig. 1.3)

- Proposed by Singer Nicolson (1972), widely accepted structure of biological membrane.
- According to this model, membranes are two dimensional solutions of oriented lipids and globular proteins.
- This model explains many cellular membrane properties such as fluidity, flexibility and ability to self-anneal and the lateral movement.

<table>
<tr><td rowspan="9">MEMBRANE TRANSPORT (Fig. 1.2)
• Simple diffusion
• Facilitative diffusion
• Active transport
• Secondary active transport
• Bulk Transport</td><td rowspan="2">Passive Transport</td><td colspan="2">Simple diffusion</td></tr>
<tr><td colspan="2">Facilitated diffusion</td></tr>
<tr><td rowspan="3">Active Transport</td><td colspan="2">ATP driven active transport</td></tr>
<tr><td rowspan="2">Ion-driven secondary active transport</td><td>Symport</td></tr>
<tr><td>Antiport</td></tr>
<tr><td rowspan="4">Bulk Transport</td><td colspan="2">Exocytosis</td></tr>
<tr><td rowspan="3">Endocytosis</td><td>Phagocytosis</td></tr>
<tr><td>Pinocytosis</td></tr>
<tr><td>Receptor Mediated endocytosis</td></tr>
</table>

Passive Transport

- It is movement of molecules across the membrane.

- It does not require energy.
- It is of two types: simple diffusion and facilitated diffusion.

Simple Diffusion	***Facilitated Diffusion***
• Uncharged or hydrophobic molecules such as O_2, CO_2, H_2O, urea and ethanol are transported by simple diffusion. • It does not involve any membrane proteins. • This process does not involve energy. • Molecules move along concentration gradient and this process is not saturable.	• It is dependent on specific integral membrane proteins. • No energy *is* used. • Molecule binds to protein on one side of membrane, undergoes a conformational change and transport molecule across the membrane and releases it on the other side. • Specific integral membrane proteins are importers that are specific for particular molecule. • They are saturable. • Example: Erythrocyte glucose transporter: Transports glucose into erythrocytes.

Ping Pong Mechanism Explains Facilitated Diffusion:

- Protein carrier in lipid bilayer associates with a solute and undergoes a conformation change (pong to ping) and releases the solute on the other side.
- Empty carrier then reverts back to original conformation (ping to pong) to complete the cycle.
- *Features:*
    - Reversible
    - Regulated by hormones

<table>
<tr><td colspan="3">Active Transport:

This type of transport requires metabolic energy which is derived from hydrolysis of ATP.</td></tr>
<tr><td rowspan="2">ATP Driven Active Transport:

• Na^+/K^+ ATPase provides energy required for the Na^+ and K^+ movement across plasma membrane.

• Also, movement of Ca^{+2}, H^+ and a no. of other molecules are coupled to hydrolysis of ATP</td><td colspan="2">Ion-Driven Secondary Active Transport:

Movement of molecule across membrane is coupled to movement of ion (usually Na^+ or H^+).

The energy for this movement is derived from movement of ion down the concentration gradient.</td></tr>
<tr><td>Symport

If molecule and ions move in the same direction, E.g. Na^+ /glucose symporter.</td><td>Antiport

If molecule and ions move in opposite direction. E.g. Erythrocyte band 3 amino transporter.</td></tr>
</table>

<table>
<tr><td rowspan="2">Glucose Rehydration Therapy

Movement of Na^+ and glucose across epithelial cells set up a osmotic pressure difference.

This results in flow of water across apical and basolateral membranes by simple diffusion.

This is the basis of glucose rehydration therapy for dehydration by administering a solution of glucose and salt to patient.</td><td colspan="2">Macromolecules are Transported Across the Plasma Membrane by Separate Mechanism of Exo-Cytosis and Endocytosis</td></tr>
<tr><td>Exocytosis

It is movement of substance out of the cell across plasma membrane.</td><td>Endocytosis

It is uptake of extracellular macromolecules cross the plasma membrane into the cell. Endocytosis is of three types namely phagocytosis, pinocytosis and receptor–mediated endocytosis.

Phagocytosis

Is ingestion of large particles such as bacteria and cell debris</td></tr>
</table>

		by forming endocytic vesicle called phagosomes, which fuse with lysosomes and is then digested by enzymes in the lysosome.

Physiologically Important Transport Systems	
Fuels:	Glucose, Fatty acid, Glutamine
Amino acids:	Most amino acids except BCAA in liver; all amino acids in muscle and some amino acid in brain; required for neural activity (glutamine, phenylalanine, tyrosine).
Ions:	Na^+, Ca^{+2}, Iron
Channelopathies:	Mutations in genes that encode for channels gives rise to disease called channelopathies. *E.g.* cystic fibrosis (Cl^- channelopathy).
CFTR	Cystic fibrosis trans membrane conductance regulator (CFTR) • Type of ABC transporter • Serves as both conductance regulator and chloride channel.

Diseases Due to Loss of Membrane Transport System	
Transport Defect	***Outcome***
Glucose-galactose Transport	Decreased uptake in intestine.
Fructose Transport System Alteration	Fructose malabsorption.
Decreased Neutral Amino Acids in Epithelial Cells	Hartnup disease symptoms in intestine and kidney.
Cystinuria	Renal absorption of cysteine and basic amino acid arginine is abnormal. Cystine renal stones. Tryptophan deficiency: pellagra-like syndrome.

Hypophosphatemia	Vitamin-D resistant rickets: renal absorption of phosphate abnormal.
Cystic Fibrosis (CF)	cAMP regulated Cl^- channel affected increasing viscosity of body secretions.

Ionophores

Certain bacteria synthesize small organic molecules, ionophores that function as shuttles for movement of ions across membranes.

They contain hydrophilic centers that bind specific ions and are surrounded by hydrophobic regions.

Each ionophore has definite ion sensitivity.

Types

1. ***<u>Mobile carriers:</u>*** They readily diffuse in a membrane and can carry an ion across the membrane.	*2.* ***<u>Channel former:</u>*** They create a channel that traverses the membrane through which ions can diffuse.

Examples

<u>Valinomycin K^+ uniport:</u>	*<u>Nigericin:</u>*	*<u>Gramicidin:</u>*	*<u>Aquaporins (AP)</u>*
1000 times more affinity for K^+ than Na^+ 10 times more affinity for Ca^{+2} than Mg^{+2}.	K^+/H^+ antiport (neutral)	H^+, Na, K^+, Rb^+ forms channels	• Proteins that form water channels in certain membrane *e.g.*, RBC, collecting ducts in kidney • Tetramer transmembrane proteins • Mutation in gene coding AP-2 causes diabetes insipidus.

Ionophores are valuable experimental tools studying ion translocation in biological membranes and manipulation of ionic composition of cells.

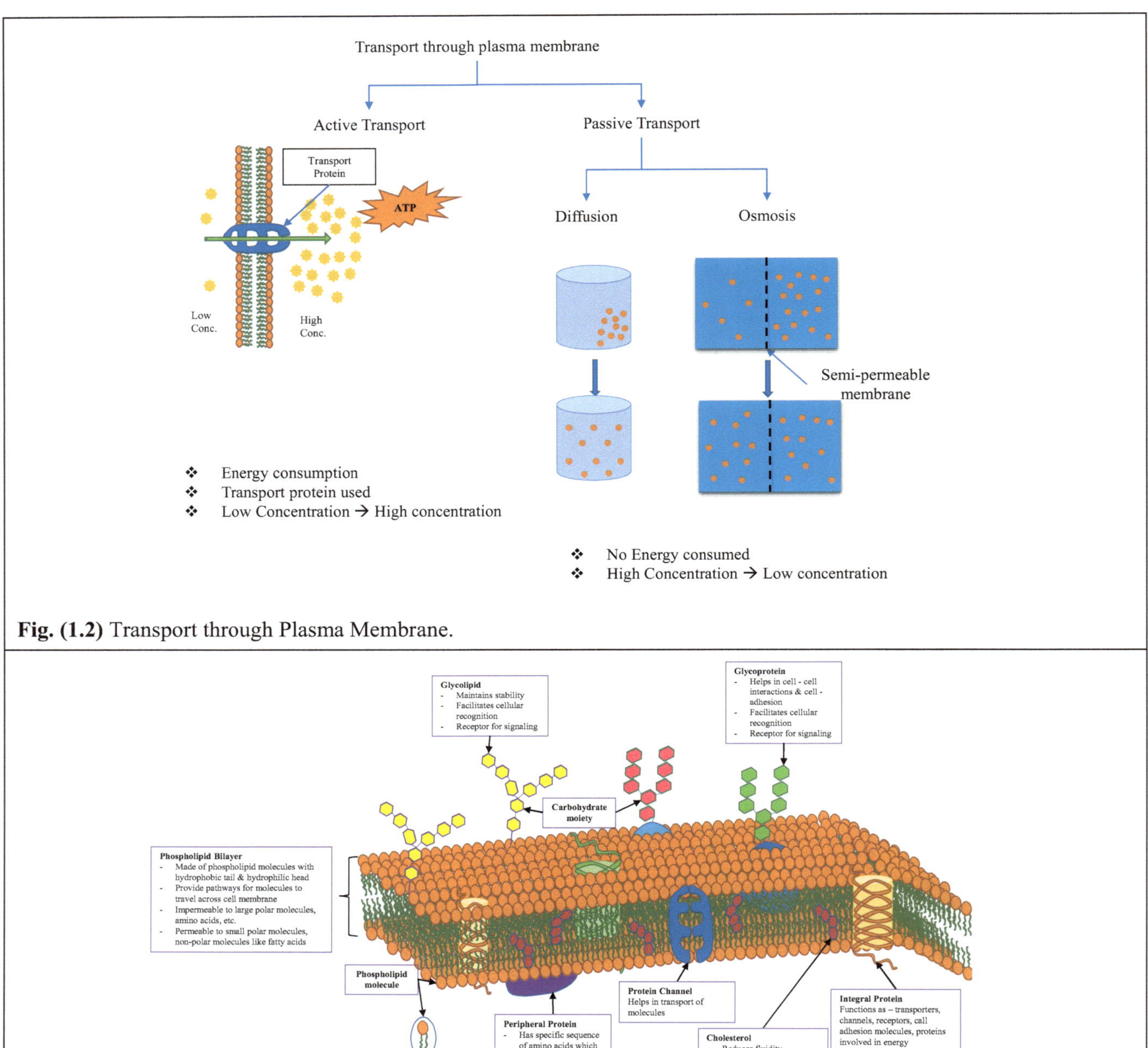

Fig. (1.2) Transport through Plasma Membrane.

Fig. (1.3). Cell Membrane.

QUESTIONS

1. Give a labelled diagram of cell.
2. Delineate various metabolic pathways in cytosol and mitochondria.
3. What are different cell junctions and their importance.
4. Importance of cytoskeleton.
5. What are different function of SER and RER.
6. Importance of cytochrome P450.
7. Clinical importance of different defect and disorders in cell organelles.
8. Difference between Lysosome and lysozyme.
9. Write short note on mitochondrial DNA.
10. Justify:
    a. mitochondria is power house of cell.
    b. lysosomes are suicide bags.
    c. importance of cytoplasmic membrane
11. Compare prokaryotic and eukaryotic cells.
12. Give an account of functions of sub-cellular organelles.
13. Describe features of biological membrane.
14. Short notes:
    a. Amphipathic nature of membrane
    b. Micelle
    c. Integral membrane proteins

d. Active transport

e. Facilitated transport

f. Fluid mosaic model.

15. Discuss types of membrane transport with suitable example.

16. State the fundamentals of cell theory.

17. Give the names of the cell components of an animal cell in the sequence in which they would be obtained as a pellet during differential centrifugation.

18. Discuss the endosymbiotic theory of origin of eukaryotic cells from prokaryotes, emphasising on the cellular compartmentalization, energy production and respiration.

19. Write the functions of the Energy-related eukaryotic cell organelles.

20. Give the similarities between *Escherichia coli* and *Euglena*.

21. What do you understand by facilitated transport across the cell membrane?

22. What is osmotic lysis of a cell? What happens to a plant cell when kept in hypotonic and hypertonic solution?

23. Differentiate between glycocalyx and peptidoglycan.

24. Differentiate between the transport of ions and water molecules across the plasma membrane.

25. How does cholesterol regulate the effect of temperature on the fluidity of cell membrane?

26. Describe the processes that move large bulk particles across the cell membrane.

27. Differentiate between osmosis and diffusion giving appropriate examples, in terms of movement of particles across a cell membrane.

28. Discuss the types and functions of membrane proteins. Describe the role of membrane proteins in maintaining cellular homeostasis.

BIBLIOGRAPHY

Bruce Alberts, Alexander Johnson, Julian Lewis, Martin Raff, Keith Roberts, and Peter Walter. Molecular Biology of the Cell. 4th edition. New York: Garland Science; 2002.

Geoffrey M. Cooper & Robert E. Hausman. The cell: A molecular approach. 7th Edition. Oxford University Press; 2019.

Victor W Rodwell, David A Bender, Kathleen M Botham, Peter J Kennelly, P Anthony Weil. Harper's illustrated biochemistry. 31st edition. New York: Mcgraw-Hill Education; 2018.

CHAPTER 2

Introduction of Metabolism: Anabolism, Catabolism and Energy Metabolism

LEARNING OBJECTIVES:	Keywords:
Explain the metabolic role of catabolic and anabolic activities in a cell. Describe the processes of how cell obtain energy to perform 'Cellular work.	ATP, NAD (P)H, Anabolism, Catabolism, Metabolic integration, Sugar phosphates.

QUICK SUMMARY	MACROMOLECULAR SYNTHESIS
Metabolism: Five Functional Blocks ***Catabolic Activities*** Foods oxidized to carbon dioxide and water; ATP and NADPH produced. ***Anabolic Activities*** Metabolic intermediates from catabolism converted to a variety of molecules; ATP and NADPH consumed.	***Anabolic Products used to Synthesize Biopolymers:*** ATP principle source of energy. GTP: Protein synthesis. CTP: Phospholipid synthesis. UTP: Polysaccharide synthesis. Photochemical activities. Light energy used to produce ATP and NADPH. Carbon dioxide fixation. ATP and NADPH used to fix carbon dioxide and convert to an intermediate.

Ten Key Intermediates	
Carbohydrates	CoA derivatives
Triose-P, tetrose-P, pentose-P, hexose-P.	Acetyl-CoA, succinyl-CoA

Simmi Kharb

α Keto acids. Pyruvate, oxaloacetate, α -ketoglutarate.	PEP. ADP/ATP and NAD/NADPH couple catabolism to anabolism.

METABOLISM FACT FILE

Intermediates: A number of intermediates that serve crucial roles in intermediary metabolism such as sugar phosphates, pyruvate, oxaloacetate, α -ketoglutarate, acetyl-CoA, succinyl- CoA and PEP.

1. ***Sugar phosphates:*** Found in glycolysis, gluconeogenesis, and the pentose phosphate pathway.

a. Pyruvate:	***b. Oxaloacetate and α -ketoglutarate:***
Derived from glycolysis and amino acids. Port of entry into the citric acid cycle for glucose-derived carbons.	Citric acid cycle intermediates. Both can be produced from amino acids by deamination.
2. Acetyl-CoA:	***3. Succinyl-CoA:***
Consumed in citric acid cycle The common denominator between fatty acids, sugars, and amino acid	A citric cycle intermediate. Place of entry of propionate from dietary sources and odd-chain fatty acid catabolism. Product of amino acid catabolism. Used in heme biosynthesis.

4. ***ATP and NADPH*:** Serve critical roles in coupling catabolism and anabolism.

Catabolism:	***Anabolic Pathways*:**
Largely oxidative in nature. Leads to a reduction of cofactors NAD^+ and FAD. ***Catabolic pathways:*** Exergonic and lead to the synthesis of ATP. ATP is then consumed in anabolic, energy requiring pathways. ***Under physiological conditions:*** ➢ Complete oxidation of glucose: • Gives high yields of ATP. ➢ The process is always far from equilibrium	Reductive with NADPH. Usually serving as an immediate source of electrons. ***NADPH:*** • This coenzyme is reduced in the pentose phosphate pathway. • Additionally, cycles exist to move electrons from NADH to $NADP^+$

ATP Equivalents

The metabolic unit of energy exchange is ATP.

Defined as the amount of energy released upon hydrolysis of ATP to ADP.

ATP Equivalent of Key Metabolic Reactions

ATP hydrolysis: 1.0.

PPi hydrolysis: 1.0.

ATP to AMP and 2 Pi, 2.0.

NADH Oxidation

3.0 ATP (2.5 in mitochondria)

FADH2 oxidation, 2.0 ATP (1.5 in mitochondria).

METABOLIC INTEGRATION

Key Features: The major organs specialize in the metabolism of particular fuels: - There is the interplay among liver, muscles, heart, adipose tissue, and brain. - This ensures that energy demands are met.	For example, glucose: Can be supplied to other tissues by: ➢ Liver by gluconeogenesis, glycogenolysis. ➢ Muscle can produce lactic acid during times of intense energy demands and this lactic acid is sent to the liver for reprocessing into glucose.
Energy demands are ultimately met by diet and humans have a complex system of hormonal regulation to regulate energy storage and appetite. Brain, stomach, small intestines, pancreas, and adipose tissue play a role in stimulating or suppressing appetite.	

BIOENERGETICS AND OXIDATIVE METABOLISM, ENZYMES

LEARNING OBJECTIVES:	Keywords:
• Describe the fundamental concepts and key mediators in the regulation of metabolic pathways. • Discuss the significance of oxidative metabolism and identify its vital constituents. • Justify the clinical correlation of metabolic pathways. • Explain the function of enzymes in biological systems. • Illustrate enzyme inhibition and its biological significance.	Bioenergetics, Diagnostic marker enzyme, Enzyme inhibition, Electron transport chain, Enzymes, Isoenzymes, Oxidative phosphorylation.

Bioenergetics and Oxidative Metabolism: Terminology

Bioenergetics	Study of energy changes occurring during biochemical reactions.
Endergonic Reaction	If a change in free energy is more than zero ($\Delta G>0$), the reaction cannot proceed spontaneously unless there is an input of energy to drive the reaction forward.
Enthalpy (H)	It is the heat content of a body.
Entropy (S)	It is the degree of randomness or disorder, of a system. Greater is degree of disorder, higher the value of S. Change in free energy $\Delta G = \Delta H - T \Delta S$ Free energy change (ΔG) of reaction is difference between total free energy of reactants and those of the products.
Exergonic Reaction	If change in free energy is less than *zero* ($\Delta G<0$), reaction proceeds spontaneously with release of energy.
Free Energy	All chemical reactions follow laws of thermodynamics (also called Gibb's free energy). Associated with every chemical compound. It is the amount of useful work that can be obtained.

<table>
<tr><td></td><td colspan="2">Free energy of compound is equal to free energy per mole time the number of moles of that component.</td></tr>
<tr><td>Standard Free Energy (G°)</td><td colspan="2">It is equal to free energy of component at unit activity level (1 mol l^{-1})
$G_1 = G_1° + RI\ In\alpha_1$
R = gas constant (1.987 x 10^{-3} kcal mol^{-1} deg^{-1})
T = temperature (°C + 273)
In = natural log (converted to log_{10} by multiplying by 2.303)
α_1 = activity of ith component.
At equilibrium, ΔG = 0, (free energy of both sides are equal)
If ΔG is negative reaction will proceed in reverse direction until equilibrium is reached.</td></tr>
<tr><td>Standard State</td><td>pH: 7
Temperature: 25°C (298°K)</td><td>Solutes at 1M (molar concentration)
Gases at 1 standard atmosphere (atm) pressure</td></tr>
<tr><td>High Energy Phosphate Compounds
All organisms obtain supply of their free energy from their environment to maintain their living processes ATP is principal donor of free energy in biological system and serves as "universal currency" for energy in cells. ATP has intermediate value of group transfer potential:
ATP → ADP + Pi; ΔG° = -7.3
ATP → AMP + PPi; ΔG° = -7.7</td><td colspan="2">High Energy Phosphates
- They include ATP, creatine phosphate.
- In humans, amount of ATP formed approximates body weight and is broken down every 24 hours.
- Creatine phosphate provides energy in muscles and is regenerated during resting phase at the expense of ATP. Other high energy phosphates include GTP, CTP, and UTP.</td></tr>
<tr><td>Oxidation – Reduction Reactions</td><td colspan="2">These reactions that involve electron transfer.</td></tr>
<tr><td>Oxidation:</td><td colspan="2">Loss of electron and reduction is gain of electrons.
A (oxidized) + B (reduced) ↔ B (oxidized) + A (reduced).</td></tr>
</table>

Free Energy and Redox Potential	Free energy changes can be expressed in terms of redox potential (E′°). $\Delta G^{\circ\prime} = -nF\Delta E_O'$. *Where F is faraday constant (23.061 Kcal x volt*$^{-1}$ *x equiv*$^{-1}$*).* *Standard free energy* can be calculated by this equation if redox potential of reaction ($\Delta E_O'$) and number of electrons transferred (n) are known. *Redox potential* (E_O) of a system is compared with potential of a hydrogen electrode and in biological system, redox potential is expressed at pH 7.0.
Oxidoreductases	Enzymes involved in oxidation and reduction Four types: Oxidase, Dehydrogenase, Hydroperoxidase and Oxygenase.

Electron Carriers	**Oxidative Phosphosporylation**
In aerobic system, molecular oxygen is the ultimate acceptor of electrons derived from the fuel molecules. Electrons are first transferred from metabolites to specialized electron carriers and then to molecular oxygen *via* mitochondrial electron transport chain. NAD, NADP, FAD, FMN are common electron carriers.	All the energy released from oxidation of carbohydrate, fat and protein is made available in mitochondria as reducing equivalents (-H or e^-) and these are funnelled into respiratory chain for form water with oxygen.

ELECTRON TRANSPORT CHAIN (ETC)	Enzymes of electron transport chain are embedded in inner mitochondrial membrane (IMM) in association with the enzymes of oxidative phosphorylation.
Overview of ETC	
1. NADH and FADH$_2$ Reduced form of NAD and FAD.	*Produced by:* Glycolysis β oxidation of fats

	TCA cycle. Pass electrons to components of ETC located in IMM. NADH freely diffuses from matrix to membrane. $FADH_2$ tightly bound to enzymes that produce it within IMM.
2. *Transfer of Electrons from NADH to Oxygen*	Occurs in three stages.
3. *Each Complex uses Energy from Electron Transfer to Proton Pump to Cytosolic Side of IMM*	
4. *Proton-Motive Force* (an electrochemical potential)	Generated and when proton enter back into matrix through ATP synthase complex and ATP is produced. Also, some energy is lost as heat during transfer of electrons through ETC.
5. *ETC has a Large Negative $\Delta G°$*	So, electrons flow from NADH ($FADH_2$) towards O_2.

Enzyme Compounds in Mitochondria			
Outer Membrane	***Inter Membrane Space***	***Inner Membrane***	***ATP Synthase***
Acyl CoA synthetase Glycerol phosphate acyl transferase	Adenylyl kinase Creatine kinase	Respiratory chain enzymes Cardiolipin	Membrane transporters

Organization of ETC: The flow of electrons through the respiratory chain generates ATP by the process of oxidative phosphorylation.

Complex I:	***Complex II:***	***Q (coenzyme Q):***
NADH-CoQ oxidoreductase Entry point for electrons from NADH. CoQ (ubiquinone) or Q is electron acceptor and FMN and Fe-S centers are prosthetic groups. Rotenone and barbiturates inhibit electron transfer from NADH to CoQ.	Succinate-Q reductase Entry point of electrons from succinate into ETC. FAD, Fe-S centers and heme (cyt b_{560}) are prosthetic groups and coenzyme Q is electron acceptor. Carboxin inhibits complex II	Accepts electrons from complex I and complex II
Complex III:	***Cytochrome C:***	***Complex IV*:**
Q-cytochrome c oxidoreductase. It passes electrons on to cytochrome c. Heme (cytochrome b and c_1), Fe-S centers are the prosthetic groups and cytochrome C is electron acceptor. Antimycin A is inhibitor of complex III.	It is the *only soluble protein* (with heme as prosthetic group) of ETC and transfers electrons from complex III to complex IV.	Cytochrome c oxidase It is electron acceptor for cyt c and its prosthetic groups are copper and cyt a, a_3. Electrons are accepted by molecular oxygen to form water electrons are transferred from cyt c to Cu^{2+}, then to cyt a, cyt a_3 and then to O_2. Carbon monoxide, azide, cyanide and hydrogen sulfide are inhibitors of electron transfer from cyt c to O_2.

Three Major Steps of ETC		
Stage I: Transfer of electrons from NADH to CoQ (complex I):	***Stage II:*** Transfer of electrons from CoQ to cyt c:	***Stage III:*** *Transfer of electrons from cyt c to cyt aa_3 (complex IV):*
- NADH produced by oxidation reaction in mitochondrial matrix diffuses to IMM and passes its electrons to FMN. - FMN passes electrons through a series of Fe-S protein complexes to CoQ. CoQ accepts one electron at a time forming semiquinone and ubiquinone. - Energy produced by these electron transfers is used to pump protons to cytosolic side of IMM.	CoQ passes electrons *via* Fe-S centers to cyt b and cyt c_1 (complex III) which transfers electrons to cyt c by cyt c reductase.	Transfer of electrons from cyt c to cyt aa_3 by cyt c oxidase and electrons are transferred to molecular O_2, reducing it to water.
	Complex II: - Electrons from $FADH_2$ enter ETC at CoQ level. - Electrons are produced by succinate dehydrogenase reaction.	*Cyt a and cyt a_3*: Heme proteins For each mole of NADH oxidized, ½ mole of O_2 converted to water. Energy produced by transfer of electrons from cyt c to O2 is used to pump protons across IMM. As electrons pass through ETC from complex I to IV, *proton-motive force is generated* which consists of both membrane potential and pH gradient. Cytosolic side has higher $[H^+]$ than matrix, proton reenter matrix through ATP synthase complex, generating ATP.

Peculiarities of Cytochrome
cyt contains heme as prosthetic group. Fe^{3+} (ferric) state of heme iron accepts one electron and gets reduced to Fe^{2+} (ferrous). Cyt carry one electron at a time and CoQ acts as an adaptor between electron transfers to complex I and III. Protons are pumped across IMM by using electron transfer from CoQ to cyt c. Proton move back *via* ATP synthase complex into matrix to drive ATP synthesis.

Fetal Infantile Mitochondria Myopathy	***MELAs***	***Leber's Hereditary Optic Neuropathy***	***Kearns-Sayre Syndrome***	***Leigh Disease***
Occurs due to decreased activity of respiratory chain complex I, III and IV. Patients present with early progressive liver failure, hypoglycemia, increased lactate in body fluids and neurological abnormalities.	Caused by mutation in mitochondrial gene encoding complex I (NADH: Ubiquinone oxidoreductase) Acronym for: mitochondrial encephalopathy, lactacidosis and stroke.	Point mutation in gene for cyt reductase Results in loss of central vision by age of 20 to 30 years.	Mutation in complex II of ETC Patients manifest with short stature, external ophthalmoplegia, retinopathy, ataxia, cardiac conduction defects.	Mutation in cyt oxidase Lactacidemia, developmental delay, hypotonia, seizures.

ATP Production	**Remember**
As electrons pass through ETC from complex I to IV, proton- motive force is generated which consists of both membrane potential and pH gradient. Cytosolic side has higher $[H^+]$ than matrix. Protons re-enter matrix through ATP synthase complex, generating ATP (Fig. **2.1**).	IMM is impermeable to protons. Fo component forms a channel through ATP synthase complex, through which protons flow. F1 is ATP synthesizing head connected to Fo stalk projecting into matrix.

Energetics	**ATP- ADP Antiport**
***For every mole of NADH oxidized*:** - ½ mole of O_2 reduced to water - 3 moles of ATP produced - For every mole of $FADH_2$ oxidized, 2 moles of ATP produced	ATP produced in mitochondria is transported to cytosol in exchange with ADP via a transport protein in IMM called *ATP-ADP antiport*er (ANT, adenine nucleotide transferase).

Inhibitors of ETC

Component	***Substance***	***Mechanism***
Complex I	Rotenone, Amytal	Inhibit NADH dehydrogenase
Complex II	Carboxin, Thenoyltrifluoroacetone	Ubiquinone type inhibitors
	Malonate	Succinate-analogue inhibitor
Complex III	Antimycin A Myxothiazol, stigmatellin	Binds to the Q_i site, inhibits the transfer of e^- Binds to the Q_o site, inhibits the transfer of e^-
Complex IV	Cyanide, azide, and carbon monoxide	Bind to cytochrome c oxidase, lead cells to chemical asphyxiation
ATP synthase	oligomycin	Binds to ATP synthase, acts as uncoupler
ATP-ADP antiporter	Atractyloside	Deplete mitochondria of ADP

Uncouplers of ETC	***Examples of Uncoupler***
Compounds that allow normal function of ETC without producing ATP. They cause leakage or transport of H+ across membranes, collapses proton gradient before it can be used for ATP synthase. Energy is released as electrons are transferred down the transport chain, but is not trapped as ATP, but released as heat.	2,4 dinitrophenol Dicumarol Chlorocarbonyl cyanide phenylhydrazone (CCCP) Bilirubin.

Oxidative Phosphorylation	It is main source of energy in aerobic cells where free energy released along ETC is coupled to formation of ATP. According to chemiosmotic theory, flow of electron and generation of ATP are coupled by a proton gradient across the inner mitochondrial membrane. This proton motive force drives the mechanism of ATP synthesis and complexes I, III and IV act as proton pumps. ATP synthase in inner mitochondrial membrane causes H^+ to pass through the membrane into matrix.
Coupling Sites for ATP Synthesis	Energy released by electron transfers catalyzed by three coupling sites: ***Complex I, III and IV*** to form 1 mole of ATP each. Electrons that enter chain from NADH form 3 mole of ATP and from FADH2 form 2 moles of ATP.
ATP Synthase:	ATP synthase consists of two major complexes known as F_0 spans the membrane F_0 and F_1. F_0 has seven subunits and F_1 has five different types of subunits, three each of α and β and one each of γ, β and ε, making nine in all. The core of F_1 complex consists of γ subunit surrounded by six α and β subunits arranged alternately. The dissipation of energy that occurs as protons pass down the concentration gradient to matrix drives the phosphorylation of ADP to ATP by the synthase.

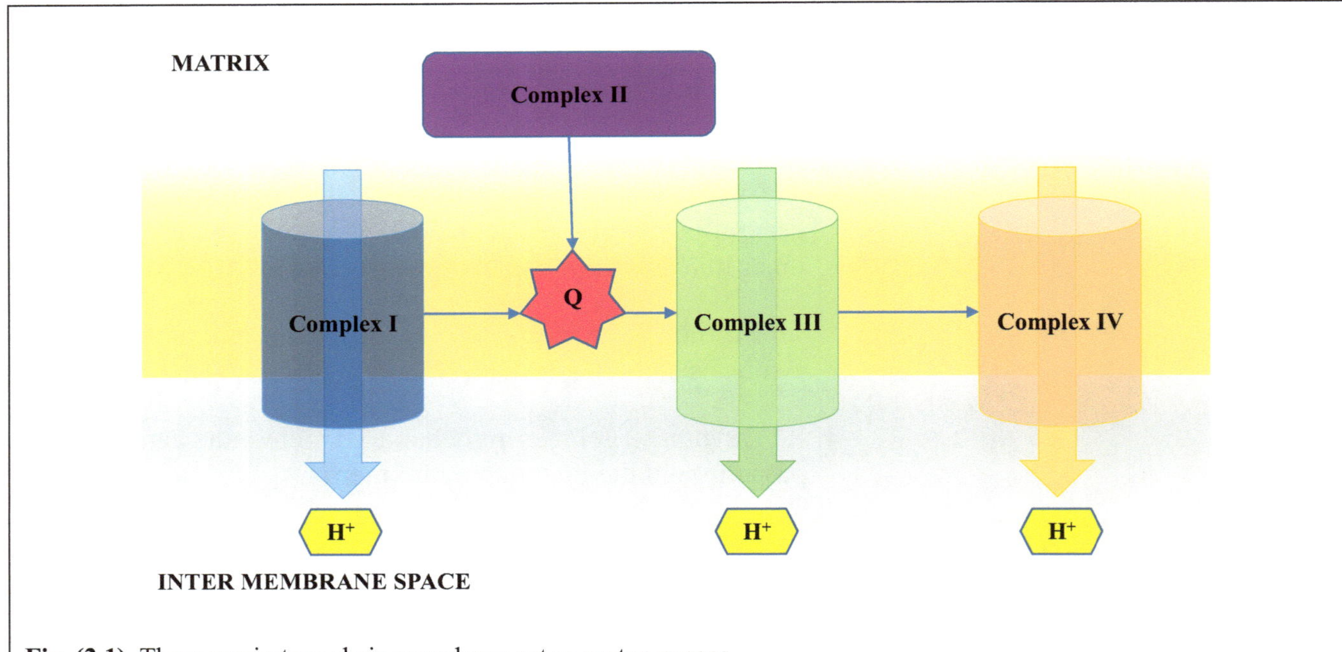

Fig. (2.1). Three respiratory chain complexes act as proton pumps.

Enzymes

Fundamental Concepts of Enzymes	***Properties of enzymes:*** Based on their active site, enzyme-substrate complex and transition state:
Enzymes: Catalysts of biological systems and most enzymes are protein in nature. Increase rate of reaction and decrease energy of activation of a reaction and speed up the reaction. Specific for substrates and products	***Active site:*** A three-dimensional structure that binds with the substrate to form enzyme- substrate complex, resulting in catalysis. ***Enzyme – substrate complex:*** Enzyme (E) binds to substrate (S) forming an enzyme – substrate complex (ES): $E + S \rightleftharpoons ES$ The formation of ES complex can occur only if the substrate possesses a group that is in correct three- dimensional orientation to interact with binding groups at the active site.

Function within a moderate pH and temperature range. Mutation in enzymes may result in many disease Enzymes are used as important diagnostic tools and therapeutic targets in a number of disease.	***Transition state:*** When a substrate is converted to a product, one stable arrangement of atoms (substrate) is converted to another (the product and this change proceeds through an unstable arrangement known as *transition state*. AB + C → *ACB* (*Transition state*) → AC + B The state of overall reaction depends on number of molecules in transition state: more the molecules in transition state, greater are the rate of reaction.

Mechanisms that Provide Enzyme Catalysis Effects 1. General acid – base catalysis 2. Covalent reactions 3. Strain within substrate	**Enzymes as Tools** *1.* ***Diagnostic tools:*** To measure the concentration of biochemical compounds, enzyme activities in blood, in forensic and genetic investigations enzymes are used. *2.* ***Therapeutic targets:*** Drugs can act on enzymes to increase or decrease their activity in many diseases and cancers.
Cofactors and Prosthetic Groups	
Coenzymes:	***Cofactors:***
Complex non -protein organic molecules that participate in catalysis by providing additional reactive groups. They have little activity in absence of enzyme. Examples: ***Vitamin B complex group:*** TPP, FMN, FAD, NAD, NADP, Lipoic acid, pyridoxal PO_4 (PLP), CoA FH_4, biotin coenzyme, methylcobalamin, deoxyadenosyl cobalamin. ***Non-vitamin coenzyme:*** ATP, CDP, UDP ***Nucleotide coenzyme***: NAD, NADP, FMN, FAD, CoA, UDPG.	Certain apoenzyme requires presence of certain metal ions for their activity which are known as *cofactors*. *Examples:* Calcium (SDH), Cobalt (arginase) Magnesium (enolase), Manganese (Cholinesterase), Chloride (Salivary amylase). ***Metalloenzymes:*** Enzymes which require metal ions for their activity and if metal ion forms an integral part of enzyme, it is called metalloenzyme. For example: Zinc (SOD), Iron (Cytochrome), Copper (tyrosinase, cyt oxidase), Molybdenum (Xanthine oxidase), and Selenium (GSHPx).

Nomenclature and Classification			
EC Number	**Enzyme class**	**Reaction**	**Example**
EC1	Oxidoreductase	Oxidation-reduction reaction (Transfer of elections/H)	LDH Alcohol dehydrogenase Xanthine oxidase
EC2	Transferase	Group transfer reaction (transfer of group other than H)	Hexokinase Transaminase
EC3	Hydrolase	Hydrolytic reaction (cleaves bond by adding water)	Pepsin Trypsin Lipase
EC4	Lyase	Non hydrolytic removal of groups to form double bond	Aldolase Fumarase
EC5	Isomerase	Isomerization Transfer of groups within a molecule	Triose phosphate isomerase
EC6	Ligase	Bond formation coupled to ATP hydrolysis	Glutamine synthetase Pyruvate carboxylase Acetyl CoA carboxylase
EC7	Translocases	Movement of ions/ molecules across membranes or their separation within membranes	Channel proteins and carrier proteins involved in transport across membrane

Enzyme Activity: It is a measure of the catalytic ability.	***Two Methods to Measure Enzyme Activity***

The activity of an enzyme varies with substrate concentration [S] and beyond maximal velocity (V_{max}) and any further increase in substrate concentration does not further increase the reaction rate.	I^{st} measures decrease in substrate concentration in a period of time. 2nd measures increase in concentration of a product after a period of time.
Michaelis and Menton Equation Michaelis and Menton derived the relation between [S] and reaction velocity (V) and proposed that: $E + S \underset{K_2}{\overset{K_1}{\rightleftharpoons}} ES \xrightarrow{K_3} EH$ Where K_1, K_2 and K_3 are rate constant and a hyperbolic equation is derived: V = Vmax[S] / Km+[S] Where Km = Michaelis constant; Km = $K_2 + K_3$ This equation is a quantitative way of describing dependence of rate of reaction on substrate concentration.	**Factors Affecting Enzyme Activity** The catalytic activity of an enzyme many be changes by many factors namely, substrate concentration, pH, temperature and inhibitors. The effects of these factors on enzyme activity and the ways by which they are studied are termed as ENZYME KINETICS. ***a. Substrate Concentration*** The activity of an enzyme varies with substrate concentration [S] and beyond maximal velocity (V_{max}) and any further increase in substrate concentration does not further increase the reaction rate.

b. Effect of pH:	***c. Temperature:***	***d. Enzyme Concentration:***
Each enzyme has an optimum pH and change in pH will alter the degree of ionization of functional, ionizable groups on enzymes. This affects the catalytic activity by modifying binding between substrate and enzyme. Most enzymes have optimum pH between 6 to 8 but exceptions are there e.g. pepsin (1-2), ALP (9-10) ACP (4-5).	Enzymes are active at optimum temperature. With increase in temperature, rate of chemical reaction increases and then falls (due to denaturation of enzyme). Exception: Thermophilic bacteria remain active at high temperatures (90° C).	Rate of reaction is proportional to the concentration of enzyme present.

e. Effect of Inhibitors ***Inhibitor:*** *It* is a compound that reduces the activity of enzyme It is usually a small molecule and can be a peptide or protein. Inhibitors can be divided into four classes:	***Reverslble*** Inhibition is reversible when the concentration of substrate is increased. Inhibitor binds either to enzyme or enzyme – substrate complex. $E + I \rightleftharpoons EI$ $E\text{-}S + I \rightleftharpoons ESI$ ESI binding relationship results in four types of responses:
Reversible Irreversible covalent Allosteric Enzyme catalyzed covalent	Competitive inhibition Uncompetitive inhibition Non-competitive inhibition Mixed inhibition

Competitive Inhibition $E + S \leftrightarrow ES \rightarrow E + P$ $E + I \leftrightarrow EI$	Inhibitor (I) competes with normal substrate for binding at active site of the enzyme and forms enzyme inhibitor complex (EI). Competitive inhibitor is structural analogue of substrate Examples: Succinate dehydrogenase inhibited by malonate.	
Type	***Mechanism of Enzyme Inhibition***	***Kinetic Effect***
Competitive	'I' competes with substrate for binding to catalytic site Inhibition is reversible by increasing substrate concentration	V_{max} unchanged K_m increased
Non-competitive	ESI complex cannot form products. Not reversed by increasing substrate concentration.	K_m unchanged V_{max} decreased
Un-competitive	ES complex formed at location other than catalytic site and inhibitor binds at different site	Apparent

		V_{max} decreased K_m decreased

Allosteric Inhibition Inhibitor binds to a specific site on enzyme that is away from the active site (distinct binding site: allosteric site). It is partially reversible.	In case of allosteric inhibition, K_m increases and V_{max} decreases. Example: ADP is allosteric activator and ATP is allosteric inhibitor of enzyme hexokinase.

Enzyme Inhibitors as Poison

Poison	*Formula*	*Example of Enzyme Inhibited*	*Action*
Arsenate	AsO_4^{3-}	Glyceraldehyde 3-phosphate dehydrogenase	Substitutes for phosphate
Iodoacetate	ICH_2COO^-	Triose phosphate dehydrogenase	Bind to cysteine SH group
Diisopropylfluorophosphate (DIFP; a nerve poison)	$F-P(=O)(-OCH(CH_3)_2)-OCH(CH_3)_2$	Acetylcholinesterase	Bind to serine OH group

Enzyme Inhibitors as Drugs

Clinical Use	*Enzyme Inhibited*	*Inhibitor*
Antibacterial	Dihydrofolate reductase	Trimethoprim, methotrexate
Antibacterial	Alanine racemase	D-cycloserine
Antifungal	Fungal squalene	Terbinafine, naftifine expoxidase
Antiviral	DNA, RNA polymerases	Cytosine arabinoside
Antiviral	Viral DNA polymerase	Acyclovir, vidarabine

Antiprotozoal	Ornithine decarboxylase	Alpha-difluoromethyl ornithine
Therapeutic Enzymes		
Type of Cancer	***Enzyme Inhibited***	***Inhibitor***
Benign prostatic hyperplasia	Steroid 5 alpha-reductase	Finasteride
Estrogen mediated breast cancer	Aromatase	Aminoglutethimide
Colorectal cancer	Thymidylate synthase	5-fluorouracil
Small-cell lung cancer, non-Hodgkin's lymphoma	Topoisomerase II	Etoposide
Hairy-cell leukemia	Adenosine-deaminase	Pentostatin
Diagnostic Importance of Enzymes		
Enzyme	***Location***	***Elevated Plasma Levels***
Acid phosphatase - ACP	Prostate	Prostatic cancer
Alkaline phosphatase – ALP	Bone, Liver	Rickets, hypoparathyroidism, osteomalacia, obstructive jaundice, cancer of bone/liver
Alanine aminotransferase – ALT	Liver (muscle, heart, kidney)	Hepatitis, jaundice, circulatory faillure with liver congestion
Aspartate aminotransferase– AST	Heart, muscle, red cells, liver	Myocardial infarction, muscle damage, anemia, hepatitis, circulatory failure with liver congestion
Amylase	Pancreas	Acute pancreatitis, peptic ulcer
γ-Glutamyl transferase – GGT	Liver, Kidney, Pancreas	Hepatitis, alcoholic liver damage, cholestasis
Creatine kinase – CK		

CK-MB	Heart	Myocardial Infarction
CK-MM	Skeletal muscle	Muscular Dystrophy
Lactate dehydrogenase – LDH		
$LDH_1 > LDH_2$	Heart, kidney	Myocardial infarction, kidney disease, megaloblastic anemia, Leukemia
LDH_2, LDH_3	blood cells	Leukemia
LDH5	Liver, Muscle	Liver disease, Muscle disease

Isoenzymes Examples ↓	Isoenzymes (or isozymes) are different forms of an enzyme which catalyze same chemical reactions but differ in physical or kinetic properties such as structure, solubility, V_{max}, K_m, optimum pH and susceptibility to inhibitors.
Lactate Dehydrogenase (LDH)	Composed of four subunits: H or M Can have four combinations: H4, A3, M, H2 M2, HM3H and M4. $LDH_2 > LDH_1$ in heart and reversal of normal pattern is seen in MI.
Creatine Kinase	Composed of two subunits: M and B and three types are present: CK_1 (BB) in brain; CK_2 (MB) in heart and CK_3 (MM) in skeletal muscle. CK_2 (N/B) levels increase with in four hours of myocardial in function.
Alkaline Phosphatase (ALP)	Different tissues have different forms of ALP such as liver, bone and placenta.

Diagnostic Importance of Enzymes			
Hepatic Disease ALT (alanine transaminase) or SGPT	***Myocardial Infarction*** CK2 (MB)	***Muscle Disease*** AST Aldolase	***Bone Disease*** (Bone isoenzyme of) ALP

ALP (alkaline phosphatase) GGT (gamma glutamyl transferase)	AST LDH	***Pancreatitis*** Amylase.	***Prostatic Cancer*** ACP (acid phosphatase) PSA (Not an enzyme)

QUESTIONS

1. Discuss classification of enzymes with suitable examples.
2. Discussion factors affecting activity of enzymes.
3. Discuss mechanism of action of enzymes
4. Short notes:

    a. Active site

    b. Enzyme inhibition

    c. Coenzyme

    d. Isoenzyme

    e. Zymogen

    f. Metalloenzymes

    g. Km

2. Discuss uses of enzymes.

3. Discuss uses of isoenzymes.

4.Write in brief about enzyme marker.

5. Write note on therapeutic and diagnostic importance of enzymes.

6. Name some heme proteins.

7. Differentiate between synthetase and synthase.

8. What is Apo enzyme
9. What is activation energy and how enzyme effect its levels?
10. What is acid base catalysis? Explain with example.
11. List enzymes requiring metal ions for its activity.
12. What is the difference between metalloenzymes and metal activating enzymes?
13. What are the different types of enzyme inhibition?
14. What is the difference between reversible and irreversible enzyme inhibition?
15. Describe the importance of Michaelis –Menton equation?
16. Enzyme activity is expressed in …………..
17. Name enzyme which is active in phosphorylated form.
18. Name some enzymes which are active in dephosphorylated form.
19. What are functional enzymes?
20. Write a note on cardiac enzymes
21. What are inducible enzymes?
22. What are isoenzymes and write their importance?
23. What are Allo-enzymes?
24. Define Allosteric regulation.
25. What are suicidal enzymes? Explain with example.
26. How covalent modification effects enzyme activity?
27. What are zymogens and their physiological importance?
28. Give two laws of thermodynamics.

29. What is free energy of a reaction?

30. Name two high energy molecules of our body and their free energy of hydrolysis.

31. Define the role of NADH in biological reactions.

32. Explain the role of coupled reactions and group transfers in driving biological reactions.

33. Explain biological energy transducers giving 2 relevant examples in details.

34. Explain the laws of bioenergetics in terms of free energy and entropy in biological reactions.

35. Name one vitamin that helps in group transfer reactions, giving appropriate examples in support of the answer.

36. Describe why ATP is an important energy transducer in biological systems.

37. Name two biomolecules that act as electron carriers

38. What is the importance of FAD?

39. Explain the electron transport chain in detail, with the help of a diagram, and describe its importance in glucose metabolism.

40. How does acetyl coenzyme A provide cofactors for the ETC? Explain the steps involved in details.

41. Describe how the electrons are transferred to oxygen in the ETC, giving the mitochondrial protein complexes and the cofactors involved.

BIBLIOGRAPHY

Bruce Alberts, Alexander Johnson, Julian Lewis, Martin Raff, Keith Roberts, and Peter Walter. Molecular Biology of the Cell. 4th edition. New York: Garland Science; 2002.

Donald Voet, Judith G Voet, Charlotte W Pratt. Fundamentals of Biochemistry. 5th Edition. New York: Wiley; 2016.

Lehninger A, Nelson D, Cox M. Lehninger principles of biochemistry. New York: Worth Publishers; 2000.

Victor W Rodwell, David A Bender, Kathleen M Botham, Peter J Kennelly, P Anthony Weil. Harper's illustrated biochemistry. 31st edition. New York: Mcgraw-Hill Education; 2018.

CHAPTER 3

Chemistry of Carbohydrates

LEARNING OBJECTIVES	Keywords:
• Describe the role of carbohydrates in biochemistry Categorize different types of sugars. • Illustrate structural formulas of different carbohydrates. • Identify and describe isomerism in carbohydrates	Aldoses, Asymmetric carbon, Disaccharide, Enantiomers, Epimers, Glycosaminoglycans, Heteropolysaccharides, Isomerism, Ketoses, Monosaccharide, Oligosaccharide, Polysaccharide, Reducing sugars

CARBOHYDRATES: STRUCTURE AND FUNCTION

Compounds having hydroxyl groups with aldehyde or keto group $(CH_2O)_n$ **Importance** Major fuel for all tissues Structural component of membrane Form carbohydrate with a specific function: Ribose (nucleotides), Galactose (lactose in milk), Glycolipid, Glycoprotein	**Classification** ***Based on no. of Sugar Units*** Monosaccharide, Disaccharide, Oligosaccharide, Polysaccharide ***Based on Functional groups**:* Aldehyde (aldose), Ketone (Ketose)

Class	Sub-class	Examples	
		Aldose	**Ketose**
Monosaccharide	Triose	Glycerol	Dihydroxy acetone

Simmi Kharb

	Tetrose Pentose Hexose	Erythrose Ribose Glucose	Erythrulose Ribulose Fructose
Disaccharide (Composed of Two Monosaccharides)	Maltose Sucrose Lactose	2 glucose units $\alpha 1 \rightarrow 4$ glycosidic bond α D glucose $+\beta$ D fructose $\alpha 1 \rightarrow \beta$ 2 bond αD galactose + βD glucose1$\rightarrow \beta$ 4 bond	Disaccharide (Composed of Two Monosaccharides)
Oligosaccharide (Composed of 2-10 Monosaccharide Units)	Malto-triose	Maltose and $\alpha 1 \rightarrow 6$ glucose	
Polysaccharide (10 or more monosaccharide unit)	-	Amylose αD: $\alpha 1 \rightarrow 4$ bond Amylopectin 24-30, monomers (αD glucose) $1 \rightarrow 4$ linkage plus $\alpha 1 \rightarrow 6$ branching	
Homopolysaccharides		Starch: $\alpha 1 \rightarrow 4$ linkage and $\alpha 1 \rightarrow 6$ branching and Glycogen: 12-14 monomers of D-glucose $\alpha 1 \rightarrow 4$ linkage and $\alpha 1 \rightarrow 6$ branching	
		Cellulose	βD glucose: $\beta 1 \rightarrow 4$ bond
		Inulin	βD fructose
		Dextran (break down product of starch)	αD glucose linked at $\alpha 1 \rightarrow 6$ bond, with few branches

ISOMERISM		
Asymmetric Carbon	Carbon atom bonded to four different atoms or groups of atoms is asymmetric carbon.	
Isomers	Presence of asymmetric carbon allows the formation of isomers. Number of isomers of a compound depends on the number of asymmetric carbon atoms (n) = 2^n *E.g.* glucose has 4 asymmetric carbon atoms, thus 2^4 = 16 isomers.	
Isomerism	***Reasoning***	***Example***
DL isomerism (stereo isomer)	The same chemical formula differs in the position of –OH group on one or more asymmetric carbon (*E.g.* C5 in glucose). Mirror images of each other.	D, L glucose
Optical isomerism (Enantiomer)	Presence of asymmetric carbon rotate plane polarized light either to right [dextrorotatory, (+)] or to left [levorotatory (-)]	Enantiomer (+) isomer (-) isomer
Epimerism	Differ in the configuration of –OH and –H glucose on C-2, 3 and 4 of glucose, galactose at C_4 and mannose at C-2 Or Conformation that differs only at one carbon atom	Mannose at C-2 Galactose at C-4
Anomerism	Differ in configuration at carbonyl or anomeric carbon α:- OH on anomeric is below the plane of the ring β:- OH is above the plane of the ring	α anomer β anomer
Aldose-ketose isomerism	Same molecular formula differ in the position of carbonyl carbon: Glucose C-1 is aldehyde; fructose C-2 is levo	Glucose and fructose

Heteropolysaccharides:	Composed of repeating units of monosaccharides their derivative or proteoglycans

GAG	*Repeating unit*	*Tissue distribution/ Function/disease*
Hyaluronic acid	GlcUA and GlcNA GlcUA $\xrightarrow{\beta 1,3}$ GlcNAc $\xrightarrow{\beta 1 \to 4}$	Synovial fluid Vitreous body Eye Cartilage Loose connective tissue Permit tumor cells to migrate through
Chondroitin	GlcUA GalNAc	Cartilage Structural function at the site of calcification in endochondral bone Diminishes in cartilage with age, contributing to the development of osteoarthritis
Dermatan sulfate	IdUA, GalNAc IdUA $\xrightarrow{\beta 1,3}$ GalNAc	Skin, valves blood vessels lung, sclera Sclera: maintain the shape of the eyeball Arterial smooth muscle cell proliferation in atherosclerotic lesions and plaque
Keratin sulfate	GlcNAc, Gal	Cornea Corneal transparency
	GlcNAc $\xrightarrow{\beta 1,3}$ Gal $\xrightarrow{\beta 1 \to 4}$ $GlcNa_2$	Loose connective tissue Ground substance Increased in cartilage with age, causes osteoarthritis
Heparan sulfate	GlcN GlcUA GlcN α1,4 GlcUA $\rightarrow$ GlcN	Lung, muscle liver, synapse, glomerulus Present on the cell surface or bound to ECM responsible for charge selectiveness Arterial smooth muscle cell proliferation in atherosclerotic lesions and plaque

Heparin	GlcN, IdUA IdUA $\xrightarrow{\alpha 1,4}$ GlcN	Granules of mast cells, liver lung, skin Function as intracellular binding site anticoagulant, bind to lipoproteins lipase causing LP release Arterial smooth muscle cell proliferation in atherosclerotic lesions and plaque
Key: GlcUA – glucuronic acid, GlcN – glucosamine, A – Acetyl		

CARBOHYDRATE DERIVATIVES

Amino sugar (Glucosamine, Galactosamine).

Sugar acids (Ascorbic acid glucuronic acid).

Deoxy sugar (2-deoxyribose).

Sugar alcohol (D-sorbitol, D-mannitol).

Phosphoric acid esters(D-glucose1-phosphate).

Glycoproteins (mucoproteins)

Carbohydrate is found attached to many globular proteins which are classed as *glycoproteins*. Glycoproteins include integral membrane proteins that function as receptors for hormones or other molecules, immunoglobulin, complement, interferon, component of mucus.

Terms

Mutarotation

α and β-forms equilibrate via the straight chain aldehyde form. This occurs due to the opening of the hemiacetal ring.

Reducing Sugar

Sugar possessing a free anomeric carbon atom that is not involved in a glycosidic linkage is *reducing sugar*.

E.g. Lactose, maltose. Sucrose is a non-reducing sugar because the glycosidic bond between anomeric carbon C1 of α glucose and C-2 of fructose are involved in bond formation.

Inversion of Sugar

Sucrose is dextrorotatory (+66.5°) and enzyme sucrase hydrolyzes it to glucose and fructose. It becomes levorotatory (-28.2°).

This process of change is optical rotation from dextro (+) to levo (-) is *inversion of sugar*.

	Glycoprotein	**Glycolipid**	**Proteoglycan**
Composition	Oligosaccharide + protein	Oligosaccharide + lipid	GAG + protein

Location	Extra cellular surface of plasma membrane, secreted proteins	Projections on membrane: cell– surface Ag recognition sites	Connective tissue joint lubricants

Organ and Site-specific Functions of GAG

GAG	*Organ/ Site*	*Function*
Heparan sulfate	Liver/ Intracellular on cell surface	Anticoagulant
Heparan sulfate	Kidney/ Renal basement membrane.	Charge selectivity of glomerulus
Keratan sulfate Dermatan Sulfate	Cornea/ In between collagen fibers	Corneal transparency
Heparin	Mast cells	Inflammatory response
Heparin sulfate	Vascular wall/-	Anticoagulant activation of lipoprotein lipase

OXIDATION OF CARBOHYDRATES	REDUCTION OF CARBOHYDRATES
Oxidized Forms (a) Gluconic acid, 6-phosphogluconate (b) Uronic acid, glucuronic acid	Aldehyde or ketone of a sugar can be reduced to a hydroxyl group, forming a polyol. Glucose is reduced to sorbitol Galactose to galactitol Sorbitol does not readily diffuse out of cells and accumulates in cells causing osmotic damage to neurons and cataract.
Tests for Reducing Sugars (Fig. 3.1) Chemical properties depend on activity of aldehyde or ketone group:	1. Reducing properties: Benedict's, Barfoed's 2. Osazone formation

	3. Furfural formation: Molisch, Seliwanoff 's

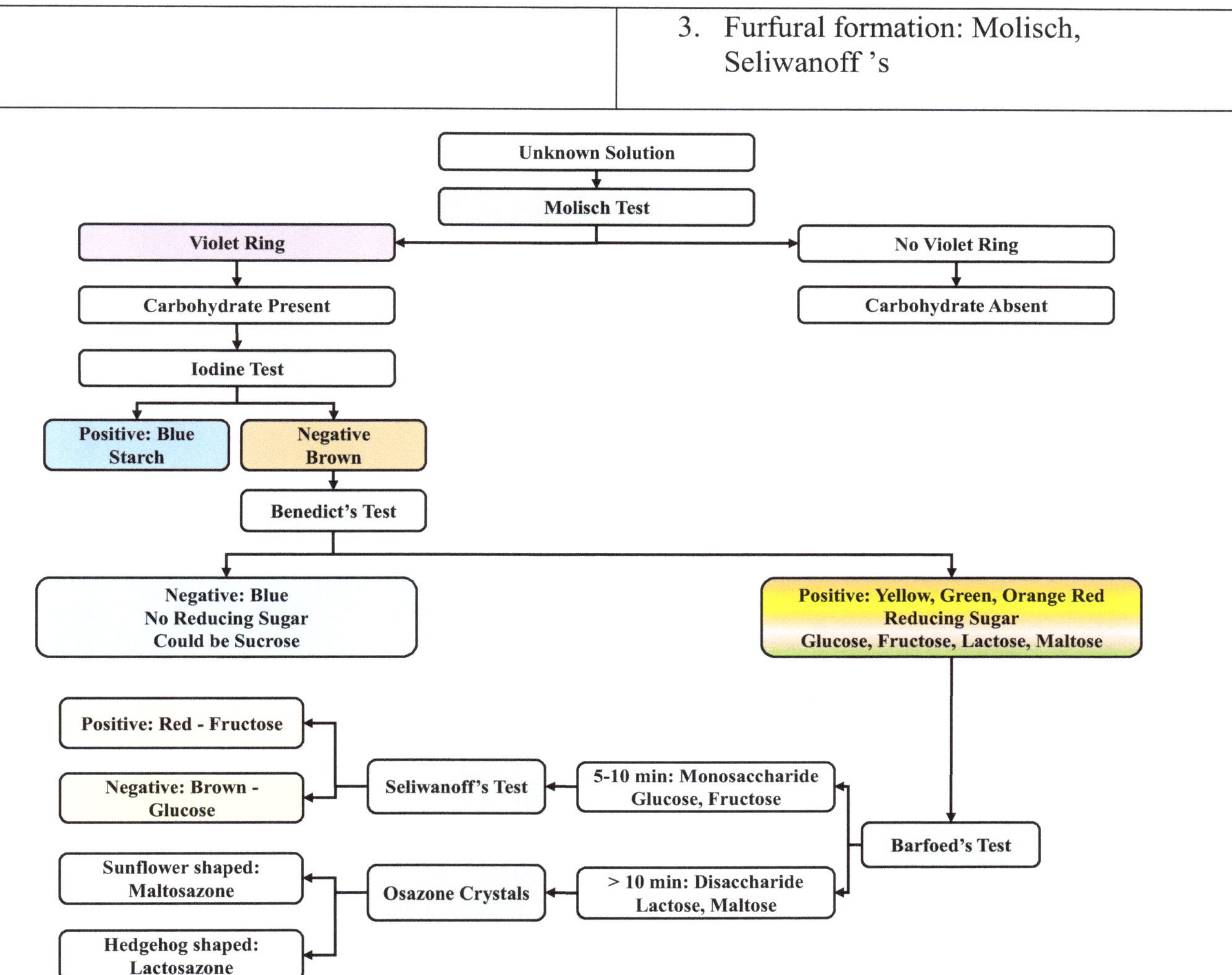

Fig. (3.1). Scheme for identification of unknown carbohydrates.

QUESTIONS

1. Classify carbohydrates with suitable examples.
2. Short notes:
    a. Reducing Sugars
    b. Asymmetric carbon

c. Mucopolysaccharides

d. Mutarotation

e. Invert Sugar

f. Anomeric carbon

g. Limit dextrin

h. Epimer

3. Write briefly:

a. Why sucrose is non-reducing sugar.

b. Difference between cellulose and starch

c. Benedict's test

d. Glycoproteins

e. Reactions of monosaccharides

f. Stereoisomerism

g. Optical isomerism

4. Classification and importance of carbohydrates.
5. What are the functions and disease associated with glycosaminoglycans.
6. What are different carbohydrate derivatives and their uses.
7. Differentiate between isomer, epimer and anomer with example.
8. What is asymmetric carbon?
9. How does mutarotation occur naturally?
10. Name biochemistry test based on reducing property of sugars.

11. What do you understand by the term inversion of sugar?

12. What is the importance of glycoproteins?

13. Classify carbohydrates on the basis of their (a) functional groups and (b) monomeric units. Name three examples of each type.

14. Describe the chemical properties of Carbohydrates, with suitable examples.

BIBLIOGRAPHY

Donald Voet, Judith G Voet, Charlotte W Pratt. Fundamentals of Biochemistry. 5th Edition. New York: Wiley; 2016.

Geoffrey L Zubay, Dennis E Vance. Principles of biochemistry. Dubuque, Iowa: William C. Brown; 1995.

Lehninger A, Nelson D, Cox M. Lehninger principles of biochemistry. New York: Worth Publishers; 2000.

Victor W Rodwell, David A Bender, Kathleen M Botham, Peter J Kennelly, P Anthony Weil. Harper's illustrated biochemistry. 31st edition. New York: Mcgraw-Hill Education; 2018.

Mind Maps in Biochemistry, 2021, 64-86

CHAPTER 4

Metabolism of Carbohydrates

LEARNING OBJECTIVES:	Keywords:
• Explain different anabolic and catabolic pathways of carbohydrate metabolism. • Describe the role of different enzymes and hormones involved in carbohydrate metabolism. • Identify the metabolic diseases related to carbohydrates.	Diabetes, Digestion and assimilation of carbohydrates, Enzymes of carbohydrate metabolism, glycolysis, Glycogen metabolism, Gluconeogenesis, Glucose homeostasis, Regulation of glucose metabolism, Tricarboxylic acid cycle.

CARBOHYDRATE STORAGE	Starch	Glycogen
Storage form of carbohydrates:	Stored in plants Both amylose and amylopectin	Stored in animals Liver and muscle cells

DIGESTION AND ASSIMILATION OF CARBOHYDRATES: Pathway for digestion of carbohydrates:

Substrate		Site	Enzyme	Products
Starch Lactose Sucrose	Fructose Glucose	Mouth ↓	Salivary amylase α1-4 glycosides	Maltose Maltotriose Limit dextrin
		Stomach ↓	NO DIGESTION	
Starch		Intestine	PANCREATIC AMYLASE (α1-4)	Maltose

Simmi Kharb

Polysaccharides	↓		Maltotriose Limit dextrin
	Epithelial brush border	**OLIGOSACCHARIDASE & DISACCHARIDASES**	
Sucrose Maltotriose Maltose α - limit dextrins	↓	Sucrase (α1-4) – Isomaltase (α1-4)	Glucose Fructose
Maltotriose Maltose		Maltase (α1-4) – Glucoamylase (α1-4)	Glucose
Lactose		Lactase (β1-4)	Glucose Galactose
	Portal vein ↓		
	Liver (metabolism) ↓		
	Circulation (glucose) ↓		
	Liver Muscle Adipose tissue		

DEFECTS IN CARBOHYDRATE ABSORPTION

S. No.	Disease	Defect	Clinical Feature
1.	Lactase deficiency	Lactose intolerance	Abdominal discomfort
		Inherited lactase deficiency	Cramps, diarrhea

		Secondary lactase deficiency	Intolerance to milk
2.	Inherited Sucrase deficiency	Sucrase deficiency	Same as lactase deficiency
3.	Disaccharidases deficiency	Disacchariduria	Fructose, sorbitol malabsorption
4.	Defect in SGLT-1	Monosaccharide malabsorption	Fructose, sorbitol malabsorption, watery diarrhea

Oral Hydration Therapy:	In cholera infection: NaCl absorption is inhibited, but not the facilitative transport of Na & glucose

ORS (oral rehydration solution) provides: →	110mM	Glucose
SGLT-1 is not inhibited and in the presence of glucose Na uptake takes place to replenish stores.	99 mM	Na^+
	74 mM	Cl^-
	39 mM	HCO^-_3
	4 mM	K^+

Enzyme	**Site of Production/ Action**	**Substrate**	**Product**
Salivary Amylase	Salivary glands/ Oral cavity	Starch	Disaccharides (maltose), oligosaccharides
Oligosaccharidases	Lining of the intestine; brush border membrane/ Small intestine	Oligosaccharides Disaccharides	Monosaccharides (*e.g.*, glucose, fructose, galactose)
Pancreatic Amylase	Pancreas/ Small intestine	Starch	Disaccharides (maltose), monosaccharides (*e.g.*, glucose, fructose, galactose)

GLUCOSE METABOLISM

Involves the following steps:	**Glycolysis:**	→ ***Importance:***
Glycolysis Krebs cycle Oxidative phosphorylation Before glycolysis begins, glucose must be transported into the cell by different glucose transporters (GLUT1-GLUT5).	A cytoplasmic pathway Converts glucose into two moles of pyruvate Releases energy captured in two substrate-level phosphorylation and one oxidation reaction.	Only pathway taking place in all the cells. Principal route for glucose metabolism The main pathway for fructose, lactose and galactose metabolism.

Glycolysis:	→ ***Provides:***
Requires 10 enzymes, 2 ATP molecules Liberates four ATP and two NADH+H^+ and pyruvate → ***Supplies precursors for other pathways*:** Pyruvate for alanine formation Dihydroxyacetone for triglyceride formation. → ***Energy source:*** The only source of energy in erythrocytes. Provides ATP to skeletal muscle in the absence of oxygen.	Pyruvate→ TCA *via* acetyl CoA G-6-P→ fatty acids *via* acetyl CoA G-6-P→ glycogen G-6-P→ PPP GA3P→ 3 carbon metabolisms DHAP→ glycerol and TAG 1,3 BPG → substrate level ATP production 2,3 BPG: *Regulates* O_2 release in RBCs Lactate→ GNG to generate glucose

→ Under anaerobic conditions: Occurs in erythrocytes, exercising skeletal muscle. It recycles NADH by making lactate to produce ATP.	

→ Reactions:

Reaction	***Input/ output***	***Clinical correlation: basis***
Preparatory phase: Energy investment *Phase I*: Glucose + ATP→ G-1-P + ADP. →+ATP→ Frutose-1,6 BP + ADP		
Reaction 1: Glucose +ATP-**GK/ HK→** G-6-P+ ADP	*Input:* glucose, 1ATP, one enzyme *Output*: one G-6P HK inhibited by G6P	MODY: maturity-onset diabetes of young. Autosomal dominant mutation in glucokinase gene Presents as non-progressive hyperglycemia.
Reaction 2: α G-6-P --**PHI**→ α F – 6 – P	-	-
Reaction 3: F-6 – P + ATP -**PFK→** F – 1, 6-BP	*Input:* 1ATP ***Committed step*** Inactive when cell [ATP] is high Inhibited by citrate Activated by F2,6 BP	Exercise intolerance

Phase II: Splitting Stage		
Reaction 4: F – 1, 6 –BP ← **aldolase** → DHAP + GA3P TPI DHAP ←——→ GA3P	*Yield*: 2 GA3P	Absence of Aldolase (in RBC, muscle): presents with hemolytic anemia TPI: hemolytic anemia
Phase III: Yield Stage		
Reaction 5: 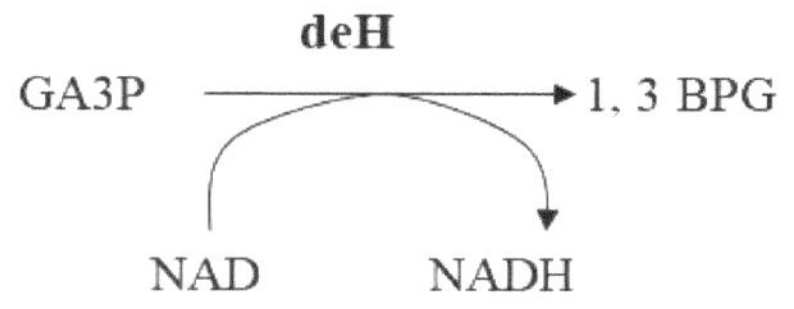 **deH** GA3P ——→ 1, 3 BPG NAD NADH	2 NADH: 6ATP	Arsenite, Mercuric ions: -SH group of lipoic acid, Arsenate used by GAPDH—> 1 arsenato-3-phosphoglycerate (instead of 1,3 BPG): it hydrolyzes and no ATP generated by substrate level phosphorylation.
Reaction 6: 1, 3 BPG <—**PGK**—> 3-PG	-	Exercise intolerance
Reaction 7: 3 PG ←**PGM**→ 2 PG	-	Exercise intolerance
Reaction 8: 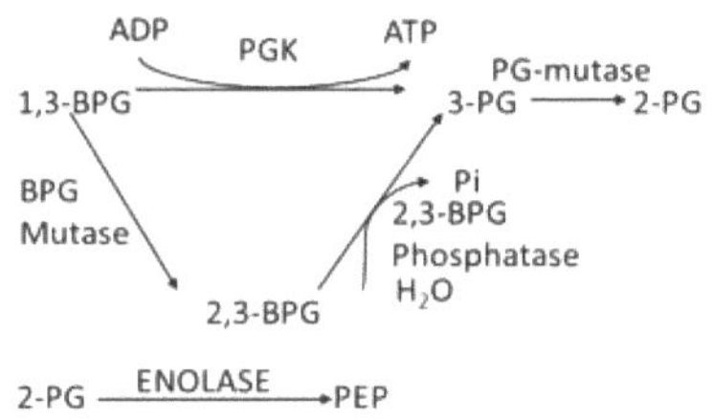 ADP PGK ATP 1,3-BPG ——→ 3-PG —PG-mutase→ 2-PG BPG Mutase 2,3-BPG Pi 2,3-BPG Phosphatase H_2O 2-PG —ENOLASE→ PEP	PKG: substrate level phosphorylation: 2 ATP	*2, 3 BPG: Adaptation to high altitude:* Increased number of RBCs Increases Hb concentration Lowers affinity of Hb for O_2 Increases ability of Hb to unload O_2 to tissues.

2PG—**enolase**—> PEP		*Enolase:* Exercise intolerance Fluoride inhibits enolase: used as preservative for blood glucose estimation
Reaction 9: PEP + ADP -- **pyruvate kinase →** pyruvate + ATP	Substrate level ATP generation Inhibited by high [ATP]	Hemolytic anemia
Reaction 10: Pyruvate —LDH→ Lactate (NAD, NADH+H⁺)	Consumption of 2NADH	Exercise intolerance Lactic acidosis
Cori Cycle: LDH enzyme oxidizes lactate to pyruvate that is used by heart as fuel.		

ENERGETICS

Reaction	**Enzyme**	**ATP gain/loss**
G→G6P	GK, HK	-1ATP
F6P→F1,6BP	PFK	-1 ATP
GA3P →1, 3BPG	GAPDH	+2NADH = +6ADP
1,3 BPG → 3BPG	PGK	+2 ATP
PEP → Pyruvate	PK	+2 ATP
Anaerobic glycolysis: Pyruvate → Lactate (LDH) 2NADH = -6ATP		Net ATP = 10-2 ATP = 8ATP

<table>
<tr><th colspan="3">Comparison of Various Pathways of Carbohydrate Metabolism</th></tr>
<tr><td>Cycle/ pathway</td><td>Features</td><td>Peculiarity</td></tr>
<tr><td><u>Glycolysis</u></td><td>Site: cytoplasm

Can occur both under aerobic and anaerobic conditions

Requirement:

10 enzymes,

2 ATP

Liberates:

four ATP and two NADH+H+ and pyruvate

Three key regulatory steps: HK, PFK and PK

<u>Energetics:</u>

Aerobic:

Four ATP

Two NADH+H+

Net gain= 8 ATP

Anaerobic: 2ATP</td><td>2, 3BPG Shunt:

2, 3BPG has a role in transport of O_2 in blood

Rate of glycolysis is higher in tumor cells

Defects:

1. Hemolytic anemia: Due to defect in Aldolase A, Pyruvate kinase, Hexokinase enzymes.

2. Exercise intolerance: Due to defect in muscle phosphofructokinase enzyme

Poisoning:

1. Arsenic, mercuric poisoning: They bind to –SH group of lipoic acid and inhibit PDH enzyme</td></tr>
<tr><td></td><td>PDH (Pyruvate dehydrogenase):</td><td>Multi-enzyme complex
Site: inner mitochondrial membrane
5 cofactors:</td></tr>
</table>

	PDH carries out oxidative decarboxylation of pyruvate to acetyl CoA	TPP, Lipoamide, FAD, Coenzyme A, NAD+ *3 enzymes*: Pyruvate dehydrogenase, Dihydrolipoyl transacetylase, Dihydrolipoyl dehydrogenase. *4 Vitamins:* Thiamin (TPP) Riboflavin (FAD) Nicotinamide (NAD+) Pantothenic acid (coenzyme A)

Cycle/ Pathway	***Features***	***Peculiarity***	***Energetics***
<u>TCA</u>	*Location:* Mitochondrial matrix, liver ***Strictly aerobic*** *Function:* production of energy, biosynthesis of glucose, FA amino acid and heme. *Eight enzymatic reactions*: four oxidations, two decarboxylation.	Common pathway for final oxidation of all metabolic fuels (carbohydrates, fats, ketone bodies and amino acids). *<u>Amphibolic:</u>* both oxidative (***catabolic***: pyruvate →acetyl CoA; TAG→ FFA+ glycerol; AA→ acetyl CoA & TCA intermediates) and synthetic (***anabolic***: Glucose, non-essential amino	Generates NADH+, $FADH_2$ and substrate level ATP One mole of glucose on complete oxidation produces 30 ATP

		acids, porphyrin: heme) process <u>*Anaplerotic reactions:*</u> Replenish intermediates of TCA cycle that participate in biosynthetic reactions Include: PEP carboxykinase, pyruvate carboxylase, transaminases, malic enzyme steps Defects: Very rare defects, usually incompatible with life.	
Cycle/ Pathway	***Features***	***Peculiarity***	***Energetics***
Glycogen metabolism	Synthesis *Site:* Liver and muscle in cytosolic fraction *REQUIREMENTS* Five *enzymes*: HK, PGM, Uridyl transferase, glycogen synthase, branching enzyme;	Readily available source of glucose between meals and during a 12-hr. fast (in liver) Products of glycogen breakdown: *Limit dextrin* (phosphorylase step)	No ATP produced

	Glycogen primer, small glycogen Two moles of *ATP*	*G-1-P* (α-1→4 hydrolysis): 90% *G-6-P* (α-1→6 hydrolysis, branch point) G-1-P is then converted to G-6-P by PGM in muscle (*** *defects)*	
	Break down: Not reversal of glycogenesis, but is a separate pathway Site: cytosol - liver and muscle	Sequential removal of glucose units from non-reducing end of glycogen	G-6-P serves as energy source for supporting muscle contraction
Cycle/ Pathway	***Features***	***Peculiarity***	***Energetics***
Gluconeo-Genesis	Biosynthesis of new glucose from various ***non-carbohydrate sources*** *Location*: Cytosolic Liver: 85-95%; ***Not in muscle*** During starvation: Kidney: 50%	Situations where GNG occurs: Normal physiological situation: Between meals, during sleep, Exercise/work, after heavy exercise or work After protein- rich diet Starvation	Does not generate ATP

	Epithelial tissue, small intestine: 5%	Metabolic acidosis, Use of dietary proteins in carbohydrate pathway.	
Cycle/ Pathway	***Features***	***Peculiarity***	***Energetics***
<u>Hmp Shunt</u>	*Location:* Cytosolic Most active in RBC, liver, adipose tissue, adrenal cortex and mammary glands. ***Does not generate ATP*** *Provides:* NADPH Pentose phosphates Alternative route for metabolism of glucose	*Maintains reduced state of Hb:* Provides NADPH to RBCs for: Reduction of oxidized glutathione (GSSG) to reduced glutathione (GSH) by enzyme glutathione reductase *Defects:* G6PD deficiency: Hemolytic anemia* Defective transketolase: Wernicke Korsakoff syndrome**	*Does not generate ATP* *Pathways requiring NADPH*: Synthesis of FA, cholesterol, neurotransmitter, nucleotide. *Detoxification:* Reduction of GSH, Cyt P450, monooxygenases
Cycle/ Pathway	***Features***	***Peculiarity***	***Energetics***
<u>Uronic Acid Pathway</u>	Alternative oxidative pathway for glucose	Defects:	*No ATP is produced*

	Occurs in liver cytoplasm Starts with G-6-P Ends with glucuronic acid, ascorbic acid and pentoses	*1. Essential pentosuria:* *G6PD deficiency* *2. Diminished activity of UDP glucuronyl transferase:* i. Neonatal jaundice: ii. Criggler Najjar syndrome iii. Gilbert's syndrome	
Cycle/ Pathway	***Features***	***Peculiarity***	***Energetics***
<u>Fructose, Galactose Metabolism</u>	Fructose: Rapidly metabolized after intake by entering glycolysis at GA3P step This bypasses two major regulatory steps of glycolysis (HK, FK) Fructokinase and aldolase B are not under control of insulin Galactose: Galactose uptake by the cells insulin – independent	Defects: *Essential fructosuria*: (Hepatic fructokinase) Hereditary fructose intolerance (Aldolase B) Fructose induced hypoglycemia (F1, 6 BPase) Fructose and sorbitol in lens cause cataract in diabetes by polyol pathway Galactose is required for synthesis of lactose, glycolipid, glycogen, glucuronic	ATP production *via* glycolysis

		acid, glycoproteins and proteoglycan Defects: Galactosemia (Galactokinase) Classical galactosemia (4-epimerase uridyltransferase)	

** In response to oxidants (such as aspirin, primaquin, sulfonamide and nitrofurans, flava beans), reduced GSH is oxidized and impairment of generation of NADPH manifests as hemolysis.*

*** Weakness or paralysis and impaired mental function.*

Defects in Glycogen Metabolism (***):

Disease	***Enzyme Defect***
von Gierk's disease or Type I	Glucose-6-phosphatase
Pompe's disease or Type II	Defective glycogen breakdown in lysosomes
Coris disease or Type III	Deficiency of debranching enzyme
Anderson's disease or Type IV	Deficiency of branching enzyme
McArdle's syndrome or Type V	Deficiency of muscle Phosphorylase
Her's disease or Type VI	Deficiency of liver phosphorylase

REGULATION OF PATHWAYS

Pathway	***Enzyme***	***Inhibitors***	***Activators***
Glycolysis	Hexokinase (HK)	G-6-P Allosteric	-

	PFK	ATP, Citrate, NADH	ADP, AMP
	Pyruvate kinase (PK)	ATP, alanine	F1, 6 BiP
TCA	Citrate synthase	*Via* feedback inhibition by ATP, Citrate, NADH, Succinyl CoA	ADP
	ICD		
	αKGDH		
	PDH SDH	acetyl CoA, NADH ↑ NADH/NAD, ↑ Acetyl CoA/CoA OAA, ↑ ATP/ADP	-
Glycogen Metabolism	Phosphorylase	*Muscle:* *Allosteric:* ATP, G-6-P *Covalent:* Insulin, ATP, Glucose *Liver:* Allosteric: Glucose	 AMP Glucagon, Epinephrine, cAMP, Ca^{+2}, AMP, Phosphorylation -

	Glycogen phosphorylase	*Covalent:* Glucagon, Epinephrine, cAMP, Ca+2, AMP, Phosphorylation	Insulin, G-6-P
Gluconeogenesis	Pyruvate carboxylase	ADP, insulin	Acetyl CoA, Glucocorticoids, glucagon, epinephrine
	Phosphoenolpyruvate carboxykinase	ADP, insulin	Acetyl CoA, Glucocorticoids, glucagon, epinephrine
	Glucose 6-phosphatase	Insulin	Glucocorticoids, glucagon, epinephrine

KREB'S CYCLE, CITRIC ACID CYCLE OR TRICARBOXYLIC ACID CYCLE (TCA)

Common pathway for final oxidation of all metabolic fuels (carbohydrates, fats, ketone bodies and amino acids). Strictly aerobic Generates NADH, $FADH_2$ and substrate level ATP. **Amphibolic**: Both oxidative (catabolic) and synthetic (anabolic) process	**Reactions:** Steps1&2: Conversion of acetyl CoA to citrate by citrate synthase; latter further converted to isocitrate *via* cis-aconitate intermediate. Step 3: Oxidative decarboxylation of isocitrate to alpha ketoglutarate by enzyme ICD. Step 4: Oxidative decarboxylation of alpha KG to succinyl CoA by enzyme alpha ketoglutarate dehydrogenase. Step 5: Substrate level phosphorylation: conversion of succinyl CoA to succinate by succinyl CoA synthase and generation of ATP.
Step 6-8: Dehydrogenation reaction: Succinate in IMM forms fumarate and NADH by dehydrogenase. Addition of water to fumarate (at trans double bond) forms malate and finally	

malate is converted to oxaloacetate by enzyme malate dehydrogenase. This completes one trip of TCA cycle.

Energetics:

Enzyme	**ATP**	
ICDH → 1 NADH	3ATP	One mole of glucose on complete oxidation produces: 2 x TCA cycle ATP = 24ATP; 2 x PDH = 6 ATP
αKGDH → 1 NADH	3ATP	
Succinate thiokinase→ 1 ATP	1 ATP	
SDH → 1 $FADH_2$	2ATP	Total = 30 ATP Net gain from glycolysis = 8 ATP
MDH → 1 NAD	3ATP	
Net ATP	12 ATP	

Anaplerotic Reactions:

These reactions replenish intermediates of TCA cycle that participate in biosynthetic reactions.

They include reactions catalyzed by PEP carboxykinase, pyruvate carboxylase, transaminases, malic enzyme.

Isoleucine, valine, methionine and threonine to succinyl CoA.

Carboxylase

Pyruvate → OAA (ATP → ADP)

PEPCK

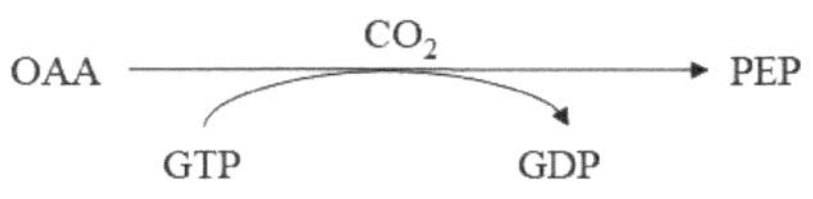

OAA → PEP (CO_2; GTP → GDP)

Malic Enzyme

Pyruvate → Malate (NAD (P) → NAD (P) H_2)

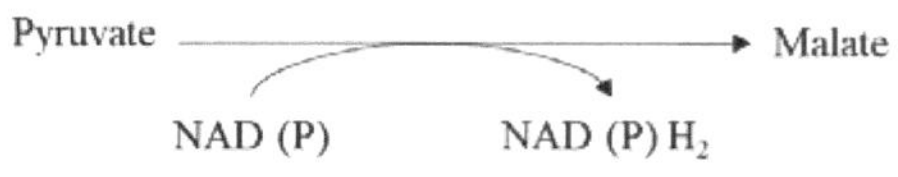

Transaminase:

GlutamateαKG →

GDH

GlutamateαKG →

Importance of regulating blood glucose levels:	*Intake:* glucose absorption from gut
Glucose is obligate fuel for CNS and RBC. Glucose turnover in a fasting 70 kg individual: 22 mg/kg/min (200 g/24hr) Plasma glucose concentration reflects balance between intake and tissue utilization.	*Tissue utilization*: glycolysis, PPP, TCA, glycogenesis. *Endogenous production*: GNG, glycogenolysis.

HORMONES REGULATING CARBOHYDRATE METABOLISM (Fig. 4.1)

	Pathway/ Step Stimulated	**Pathway / Step Inhibited**
Insulin	Glycogen synthase Glycolysis	Synthesis of enzymes of GNG
Glucagon	Induces GNG enzymes Glycogenolysis	Pyruvate kinase Glycogen synthesis
Cortisol	Induces all enzymes of GNG Amino acid degradation	-

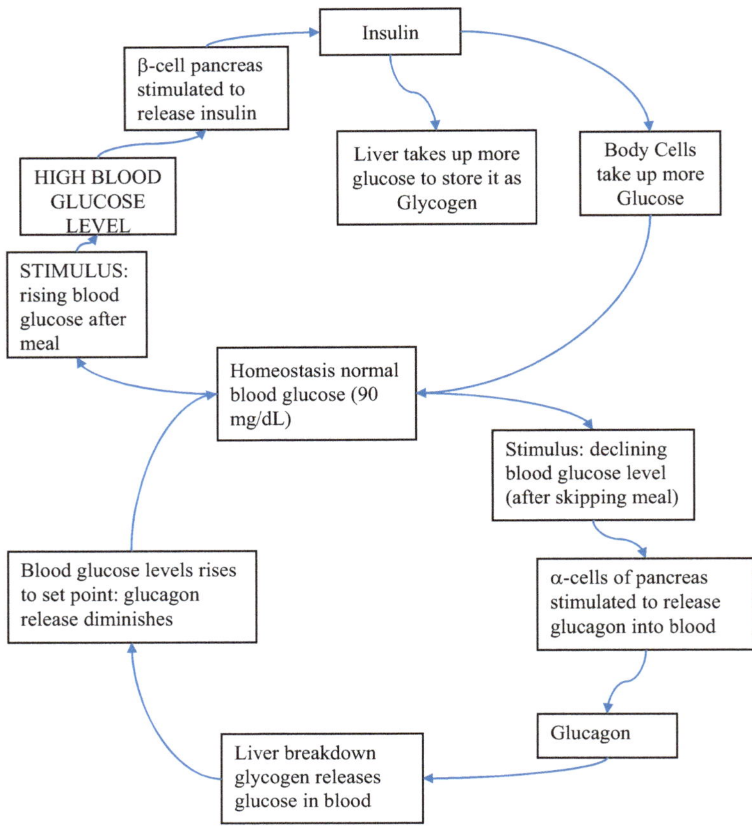

Fig. (4.1). Glucose Homeostasis.

Diabetes	**Type 1 or Insulin – Dependent Diabetes Mellitus (1DDM)**	**Type 2 or Non-Insulin Dependent DM (NIDDM)**
Features of diabetes mellitus	***Type 1***	***Type II***
Other names	Juvenile onset Insulin dependent (IDDM)	Maturity onset, Non-insulin dependent (NIDDM)
Defect	Autoimmune destruction of β-cells	Defective insulin secretion and insulin resistance

Age of onset	6 months-25 years	>40 years
Body physique	Lean	Obese
Prevalence	Lower	Higher
Treatment	Insulin	Diet, Drug, Insulin

Lab Features of Diabetes

Normal blood glucose: 80-90mg/dL

Diabetic patient: 110-140 mg/dL

After Meals:

Normal 120-140 mg/dL

Diabetic patient >200mg/dL

Symptoms

Hyperglycemia (high blood glucose)

Glycosuria (glucose in urine)

Polyuria (passage of copious urine)

Polydypsia (drinking large amount of water)

Polyphagia (increased appetite)

Weight loss

Acids and ketones in blood from lipid breakdown

Coma, if ketones build up

Complications of Diabetes

1. Reduction in blood flow to feet causes tissue death, ulceration, infection, loss of toe or feet.
2. Nephropathy – kidney damage.
3. Retinopathy – damage to blood vessel of retina cataract, skin infection, periodontitis, and neuritis.

Diagnostic Tests Include

Blood sugar, Self-monitoring of blood glucose (SMBG), Fasting plasma glucose (FPG), Fasting blood sugar (FBS), Fasting blood glucose (FBG), Glucose challenge test, Oral Glucose tolerance test (OGTT).

QUESTIONS

1. Describe digestion and absorption of carbohydrates
2. Discuss briefly:
    a. Glucose transporters
    b. Amylase
    c. Lactose intolerance
3. Discuss glycolysis along with energetics.
4. Discuss importance of HMP shunt.
5. Describe citric acid cycle.
6. Compare glycolysis and gluconeogenesis.
7. What is gluconeogenesis? Give an account of conditions favoring gluconeogenesis.
8. Explain the process of glycogen synthesis.
9. Discuss difference in glycogen metabolism in muscle and liver.
10. Discuss briefly:
    a. Glucose – alanine cycle
    b. Von Gierke's disease
    c. Glycated haemoglobin
    d. Glucose tolerance test
    e. Insulin actions
    f. Glucagon
    g. Diabetic keto acidosis

   h. Complications of diabetes
   i. G6PD deficiency
   j. Flavism
   k. Galactoseomia
   l. Fructose intolerance.

11. Discuss hormonal regulation of blood glucose levels.
12. Discuss diabetes mellitus.
13. What is the end product formed after action of salivary amylase on starch?
14. Name various disorder associated with defect in enzymes of carbohydrate digestion.
15. What is lactose intolerance? How will you diagnose it?
16. What are the conditions when glycolysis is altered?
17. Write short note on 2,3 BPG pathway.
18. Name disorders of glycolysis.
19. Write a brief note on regulation of glycolysis.
20. Differentiate between glycerol phosphate shuttle and malate aspartate shuttle.
21. How does the metabolism of RBC differ from other tissues?
22. What is Cori cycle?
23. What are the different components of PDH complex?
24. Give example of substrate level phosphorylation.
25. Describe the amphibolic nature of TCA cycle.
26. Why is glucose stored as glycogen?

27. What is the difference between glycogenesis and glycogenolysis?

28. What are the end products of glycogen breakdown?

BIBLIOGRAPHY

Denise R Ferrier. Lippincott illustrated reviews: biochemistry. 7th Edition. Philadelphia Wolters Kluwer; 2017

Donald Voet, Judith G Voet, Charlotte W Pratt. Fundamentals of Biochemistry. 5th Edition. New York: Wiley; 2016.

Lehninger A, Nelson D, Cox M. Lehninger principles of biochemistry. New York: Worth Publishers; 2000.

Victor W Rodwell, David A Bender, Kathleen M Botham, Peter J Kennelly, P Anthony Weil. Harper's illustrated biochemistry. 31st edition. New York: Mcgraw-Hill Education; 2018.

CHAPTER 5

Chemistry of Lipids

LEARNING OBJECTIVES	Keywords
• Identify the structure and name the lipids. • Classify the types of lipids. • Understand the biological role of lipids.	Ceramides, Eicosanoids, Fatty acids, Glycosphingolipids, Phospholipids, Saturated and unsaturated fats, Sphingolipids, Sphingomyelin, Triacylglycerols.

Lipids

- Hydrophobic, Insoluble in water-soluble in polar solvents.
- Composed of saturated or unsaturated long-chain hydrocarbons with a carboxyl group at end of chains.
- Important dietary constituent.
- Serve as a source of energy, thermal insulator, an important component of cell membrane, lipoprotein serves transport function, storage for TAG, stored in adipose tissue, precursor for steroid hormones.

Fatty Acids (FA)

Exist as free or esterified to glycerol

General Formula

Saturated fatty acid:

$CH_3 - [CH_2]_n - COOH$

n = number of methylene groups

Nomenclature

Systemic name: number of carbons

Saturated FA are named by chain length

Unsaturated FA are named by position of double bonds

Suffix - *anoic,* followed by acid

- *anoic* for saturated FA

- *enoic* for unsaturated FA

Simmi Kharb

<table>
<tr><td colspan="2">Position of Double Bond

Δ^9: Double bond between C9 and 10 from the carboxylic end.

ω^6: Double bond on sixth carbon from the omega end.</td><td>E.g. Palmitic acid, 16C: hexadecanoic acid.</td></tr>
<tr><td colspan="2">Naming
Carbons are numbered from carboxyl carbon (carbon no. 1).
Carbon adjacent to it is carbon number 2 (∝-carbon).
Terminal methyl carbon: ω-carbon or n-carbon.</td><td rowspan="2">Delta Numbering System
Based on the number of carbons, the number of double bonds, and position of double bonds
E.g., linoleic acid is designated as 18: 2: $\Delta^{9, 12}$

(18 carbons, two double bonds, and position of the double bond after C9 and C12 from carboxyl end).</td></tr>
<tr><td colspan="2">Classification</td></tr>
<tr><td colspan="2">Simple Lipids — Esters of FA with alcohol:</td><td>Complex lipids —Esters of FA with alcohol and contain an additional group:</td></tr>
<tr><td>Fats—esters of FA with glycerol.</td><td>Waxes—esters of FA with higher molecular weight alcohols.</td><td rowspan="2"><u>a. Phospholipids</u>

Phosphatidic acid: Esters of FA with alcohol and phosphoric acid residue.

-If alcohol is glycerol: glycerophospholipids (GPL).

-If alcohol is sphingosine: sphingophospholipids.</td></tr>
<tr><td colspan="2">Precursor and Derived Lipids

Include FA, glycerol, steroids, and alcohol.

In addition, derived lipids include ketone bodies, steroids, fatty aldehydes,

Prostaglandins, lipid soluble vitamins, and hormones.

Neutral Lipids.</td></tr>
</table>

b. Glycolipids

Glycolipids (glycosphingolipids [GSL]):

-Esters of FA, with sphingosine and a carbohydrate.

c. Other Complex Lipids

Sulfolipids, aminolipids, and lipoproteins.

Phosphatidic acid: glycerol + 2 acyl residues +PO_4.

Systemic Names of Lipids

Name	***No. Of Carbon Atom***	***Systemic Name***	***Double***
Lauric acid	12	Dodecanoic acid	-
Myristic acid	14	Tetradecanoic acid	-
Palmitic acid	16	Hexadecanoic acid	-
Stearic acid	18	Octadecanoic acid	-
Palmitoleic acid	16	Cis Hexadecenoic acid	1:9 (ω9)
Oleic acid	18	Cis. Octadecenoic acid	1:9 (ω9)
Elaidic acid	18	Trans-octadecenoic acid	1:9 (ω9)
Linoleic acid	18	Cis-9, 12 Octadecadienoic acid	2:9, 12 (ω6)
Linolenic acid	18	Cis-9, 12, 15 Octadecatrienoic acid	3:9, 12, 15 (ω3)
Arachidonic acid	20	Cis 5, 8, 11, 14 Eicosatetraenoic acid	4:5, 8, 11, 14 (ω6)

Unsaturated Fatty Acid

May contain one or more double bonds

Monounsaturated	One double bond

Isomerism in Unsaturated Fatty Acid

Naturally occurring unsaturated FA has is double bonds.

Polyunsaturated	Two or more double bond	Geometric isomerism R H R H C C ‖ ‖ C C R H H R (Cis) (Cis) (Trans) (Trans)
Eicosanoids	Prostanoid, leukotriene, prostacyclin, thromboxane	
Cis Configuration Molecule bent at 120° in double bond and produces *kinks*.	***Trans Configuration*** Molecule remains straight at the double bond.	With an increase in chain length: - Melting point (M.P.) of even-numbered fatty acids increases - M.P. decreases with instauration

TRIACYLGLYCEROL (Fig. 5.1)

- Ester of fatty acid and glycerol
- Naturally occurring fat, storage form of fat

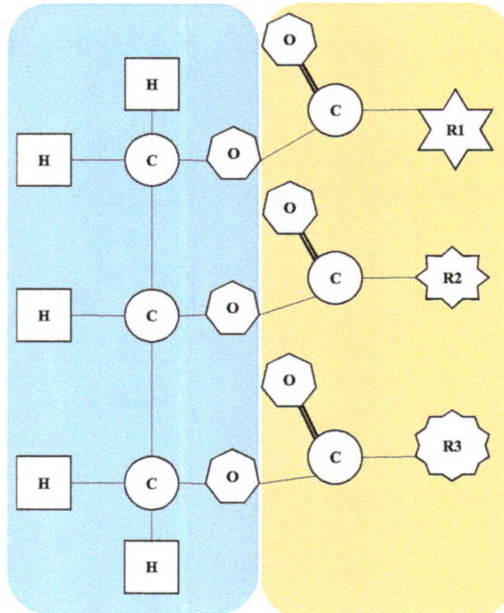

Fig. (5.1). Triacylglycerol.

PHOSPHOLIPID (Fig. 5.2)

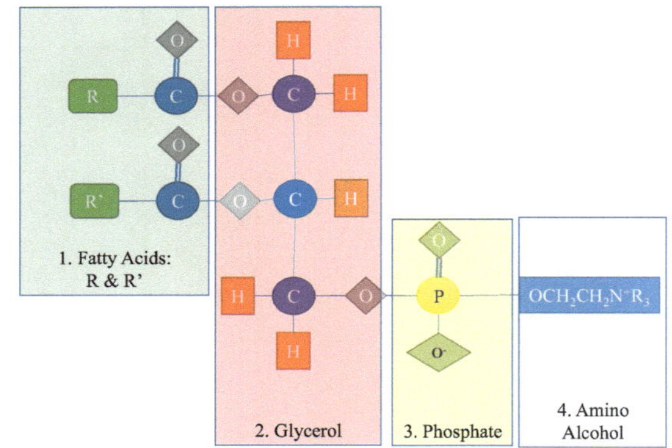

Fig. (5.2). Phospholipid.

1. Fatty acid
2. Platform back bone to with FA attached
3. PO_4
4. Alcohol attached to PO_4

R_1 and R_3: saturated FA R_2 unsaturated	

Glycerophospholipids (GPL) Major class of membrane lipids Abundant in all biological membrane. Simplest GPL is phosphatidic acid or diacylglycerol GPL includes:	GPL: Amphipathic Contain: ✓ Phosphorylated head ✓ 3-C glycerol back bone ✓ 2 hydrocarbon fatty acid chains
Phosphatidylcholine (PC) Phosphatidylcholine ethanolamine (PE) Phosphatidylcholine serine Phosphatidylcholine glycerol Phosphatidylcholine inositol Diphosphatidyl glycerol (Cardiolipin).	On C – 1 = saturated FA is found: *e.g.* palmitic, stearic acid On C- 2 = unsaturated FA is present *e.g.* oleic, linoleic, linolenic acid Glycerol ether phospholipids α - β Unsaturated. Plasmalogen is ethanolamine ether/ choline esterified to PO_4.
Sphingo Lipids In sphingomyelin, terminal OH group of sphingosines is esterified to Phosphoryl choline 2nd major class Contains sphingosine back bone rather than glycerol	*Sphingomyelin* has structural similarity to GPL and have many properties in common: *E.g.* Sphingomyelin are amphipathic with charged head group. Glycosphingolipids, sphingomyelin are classified phospholipids.

<table>
<tr><td colspan="2">Sphingosine is linked to FA by amide bond forming CERAMIDE</td><td></td></tr>
<tr><td colspan="2">Glycosphingolipids:

Have sugar moiety or primary: OH group of sphingosine (in ceramide). Examples:</td><td>Sugar containing lipids built on backbone ceramide

Lack PO_4</td></tr>
<tr><td>Glucocerebroside</td><td colspan="2">present in non-neuronal tissue</td></tr>
<tr><td>Galactocerebroside</td><td colspan="2">present in brain and nervous tissue</td></tr>
<tr><td>Ganglioside</td><td colspan="2">contain sialic acid residue in head groups.

Present in brain (represent 5-8% of total lipid in brain)</td></tr>
</table>

QUESTIONS

1. Classify lipids with suitable examples.
2. Describe briefly:
   a. Structure of cholesterol, and its functions
   b. Essential fatty acids
   c. Phospholipids.
   d. Triacylglycerol (triglycerides)
   e. Gangliosides
   f. Cerebrosides
3. What are lipoproteins, give their functions.
4. Short note:
   a. HDL cholesterol

   b. Trans fats

   c. Rancidity

   d. Bad cholesterol

5. Importance of unsaturated fatty acids

6. What is ceramide?

7. Discuss the types and functions of lipids found in the cell membrane.

8. Differentiate between PUFAs and trans-fatty acids.

9. Describe the life style changes that should be followed by a person suffering with obesity.

10. What is the role of fatty acids in prevention and development of heart disease?

BIBLIOGRAPHY

Denise R Ferrier. Lippincott illustrated reviews: biochemistry. 7th Edition. Philadelphia Wolters Kluwer; 2017

Donald Voet, Judith G Voet, Charlotte W Pratt. Fundamentals of Biochemistry. 5th Edition. New York: Wiley; 2016.

Geoffrey M. Cooper & Robert E. Hausman. The cell: A molecular approach. 7th Edition. Oxford University Press; 2019.

Jeremy M Berg, Gregory J Jr Gatto, Lubert Stryer, John L Tymoczko. Biochemistry. 9th Edition. New York: Macmillan International Higher Education: WH Freeman; 2019.

Lehninger A, Nelson D, Cox M. Lehninger principles of biochemistry. New York: Worth Publishers; 2000.

CHAPTER 6

Metabolism of Lipids

LEARNING OBJECTIVES:	Keywords:
• Illustrate the assimilation of lipids by the human body. • Explain fatty acid synthesis and degradation. • Explain the metabolism of cholesterol, eicosanoids, ketone bodies and lipoproteins. • Appraise clinical correlation of lipid metabolic disorders.	Alpha and omega oxidation, Beta-oxidation, Cholesterol, Eicosanoids, Ketone bodies, Lipid absorption and digestion, Lipid metabolism syndromes, Lipoproteins, Saturated and unsaturated fatty acid synthesis.

DIGESTION AND ABSORPTION OF LIPID	**Fact file:** Adult human digests 60-150g lipid/day and TAG constitute >90% of intake
Difficulty in the Absorption of Lipids	***Solution***
Lipids are hydrophobic TAG too large to be absorbed Digestive enzymes are water soluble	Lipids (including cholesterol) fats absorbed dissolved in lipid micelle Digestive enzymes act at water lipid interfaces

Digestion of Fats: Steps of Digestion

	Location	*Step*	*Enzyme*
1.	*Mouth, stomach: Minor*	TAG → DAG+FFA	Lingual/gastric *lipase*
2.	*Small intestine: Major*	TAG → MAG + FFA CE → Cholesterol + ester PL → FA + LysoPL	Pancreatic *lipase* *Cholesterol esterase* *PLA_2*
3.	*Small intestine lumen:*	Formation of mixed micelle	-

Simmi Kharb

4.	*Intestinal epithelial cells:*	Passive absorption of lipolytic products	-
5.	*Lymphatics:*	Assembly and export of chylomicrons	-

Lipid Absorption:

Site	***Reaction***	***Enzyme***	
Stomach: Gastric lipase	TG ↓ diacyl glycerol+ FA (MCFA)	CE	PL
Small Intestine:	↓	↓	↓
Pancreatic lipase	Sn-1 Sn-2	Cholesterol esterase	PLA_2
Colipase	→	↓	↓
Phospholipid	2MAG	Cholesterol	sn_2 FA
Bile acid	→	+	+
Micelle	IMAG	FA	Lysophosphatidylcholine
	↓ Micelle	bile acid FA PL MAG } →	Enteric brush border

Lipid Malabsorption

Defect: due to defective intestinal lipolysis or defective mucosal cell metabolism.	Features: Steatorrhea, loss of fat-soluble vitamins in stools.	Causes: Tropical sprue, Non-tropical sprue, Celiac sprue, intestinal lipodystrophy (Whipple's disease).

METABOLISM: Comparison		
	FA Synthesis	**FA Degradation (Oxidation)**
Intermediates	Linked to –SH in proteins (acyl carrier proteins)	Linked to CoASH
Site	Cytosol	Mitochondria
Enzymes	Components of a single peptide	Separate polypeptides
Reducing equivalents	$NADP^+$ / NADPH	NAD^+/NADH
Energy	Requires ATP	Produces ATP
Starts at	Carboxyl end	Methyl end
Carrier	Acyl Carrier protein	CoA
Activation by citrate	Yes	No
Inhibited by palmitate	Yes	No
Acyl/acetyl group carrier	Citrate (into cytosol)	Carnitine (cytosol to mitochondria)
Product	Palmitate	Acetyl CoA
Higher activity	Fed state	Fasting, starvation
Insulin/glucagon	High	Low
Malonyl CoA	Source of two carbon	Not involved

KEY FEATURES OF LIPID METABOLISM

Fat Oxidation:

Applied Aspect	***Features***	***Peculiarity***	***Energetics***
MCAD Deficiency Carnitine deficiency, deficiency of enzymes in fatty acid oxidation Inhibition of fatty acid oxidation occurs by poisons: hypoglycin Dicarboxylic aciduria ***Jamaican Vomiting Sickness*** Caused by eating of unripe fruit of the *akee* tree, which contains toxin hypoglycin This inactivates MCADH and SCADH, inhibiting β-oxidation ***Zellweger Syndrome*** Accumulation of VLCFA because peroxisomes are not properly formed, C26-C68 polyenoic acids accumulate in the brain. ***Adrenoleukodystrophy*** Due to the inability to transport VLCFA into peroxisomes	Oxidization of FA to acetyl CoA ***Types*** Alpha, beta, omega	Fatty acid oxidation is active in: All cells and tissues except RBCs and brain ***Reason*** *RBCs* lack mitochondria Brain/neuronal cells do have mitochondria, but there is only limited transport of FA across the blood-brain barrier ***Causes of Increased Fatty Acid Oxidation*** *Occurs in:* Starvation Diabetes mellitus Ketone body production ***Ketoacidosis*** Production of ketone bodies in excess occurs in diabetes → *Impairment in fatty acid oxidation leads to hypoglycemia:*	

Refsum Disease Deficiency of α hydroxylase Phytanic acid accumulates in tissues		Because GNG is dependent on fatty acid oxidation	

Beta Oxidation		
Features	***Peculiarity***	***Energetics***
Location: Mitochondria of all tissues ***Strictly Aerobic*** Four repeat reactions for every two carbons ***Connection:*** Fatty acid oxidation results in: Acetyl CoA $FADH_2$ $NADH_2$ (from each round of oxidation) ***Steps:*** Fatty acyl CoA released in mitochondrial matrix undergoes *beta-oxidation via* a cycle of *four back-bone reaction sequence* namely: Dehydrogenation Hydration Dehydrogenation Thiolysis	Oxidation of LCFA can occur mitochondrial membrane very slowly ***Carnitine Shuttle*** transfers FA to mitochondria for their further beta-oxidation ***Rounds of Reactions:*** Cleaves two carbons at a time from carboxyl end: release one acetyl CoA per cycle Oxidation of *odd chain FA* yields *propionyl CoA* which is converted to succinyl CoA, a constituent of TCA cycle	Each cycle of beta-oxidation reduces the length of the fatty acid chain by 2 carbons to produce: 1 acetyl CoA (12 ATP) 1 FADH2 (2 ATP) 1 NADH (3 ATP) ***Total ATP produced= 96 + 35 = 131*** Two ATP are used= 131 -2 For activation= 129 Total ATP produced= 96 + 35 = 131 Two ATP are used = 131 -2 *Final yield= 129*

Peroxisomal Beta-oxidation	***Functions***	Dehydrogenation in peroxisomes not linked directly to phosphorylation and generation of ATP
β-oxidation confined to FA more than 20C long: Does not attack SCFA Ends at octanoyl CoA (C8)	*Retailoring* of VLCFA, C-27 bile acid intermediates *Degrading* LCFA and branch chained FA Prostanoid synthesis	
β-oxidation of Odd Chain FA Odd chain FA are degraded to finally release acetyl CoA and propionyl CoA	*Propionyl CoA* gets converted into succinyl CoA which enters Kreb's cycle. *Enzymes:* dependent on coenzyme *biotin and* B_{12} (5′-deoxyadenosylcobalamin) *deficiency* of which results in methyl malonyluria.	Incorporation of the Succinyl CoA originated from the Propionyl CoA, has *generated 6 additional ATPs*: The sequence of reaction requires the consumption of 1 ATP. *Net gain*= + 5 ATP to the previous calculations
Alpha Oxidation:		
Features	***Peculiarity***	***Energetics***
Occurs in microsomes. Only *one carbon is removed* at the carboxyl end.	It requires monooxygenases, ascorbic acid and Fe^{+2} to produce α-hydroxy acid. α-oxidation of brain cerebrosides and other sphingolipids containing α hydroxyl FA. α-oxidation in liver and kidney is important for breakdown of branched-chain FA.	It is not linked to high energy production

Omega Oxidation		
Features	***Peculiarity***	***Energetics***
Occurs in endoplasmic reticulum. Involved in metabolism of drug molecules	Requires cyt P450, oxygen and NADPH to form dicarboxylic acid. Omega (last) carbon with respect to carboxyl end is oxidized to corresponding alcohol to form dicarboxylic acid	It is **not linked** to high energy production

Fat Synthesis:			
Applied Aspect	***Features***	***Peculiarity***	***Energetics***
MCAD Deficiency Medium chain fatty acyl dehydrogenase deficiency	***Location:*** Cytosol ***Organs:*** Liver, kidney, mammary gland, adipose tissue, brain ***Substrate***: Acetyl CoA ***Cofactors:*** NADPH, ATP, Biotin, Mn^{+2}, HCO_3 ***End product:*** palmitate	Human can synthesize all the FA but two: linoleic and linolenic acid which are essential fatty acid, and are required in diet Reactions of fatty acid synthesis all occur on one enzyme: fatty acid synthase (FAS) to yield free fatty acid (FFA) FAS enzyme: multienzyme complex: dimer of two identical polymonomers	*Via* citrate: 8 acetyl CoA (mitochondria to cytosol) = - 16 ATP 7 acetyl CoA → malonyl CoA = **-7 ATP** 14 NADPH → 14 NADP = -42 ATP ***Total cost = 65 ATP for C_{16}***

Saturated FA Synthesis	
Features	***Peculiarity***
Location	All reactions occur on one enzyme: fatty acid synthase (FAS): multienzyme complex: dimer

Cytosol ***Organs:*** Liver, kidney, mammary gland, adipose tissue, brain ***Substrate:*** Acetyl CoA ***Cofactors:*** NADPH, ATP, Biotin, Mn+2, HCO3 ***End product:*** Palmitate	Dimer with six enzyme activities: AT Acetyl transacylase MT Malonyl transacylase CE condensing transacylase TE Thioesterase DH dehydratase ER Enoyl reductase KR keto acyl reductase Each single polypeptide monomer has with 3 distinct domains: Domain 1: Substrate entry and condensation unit AT, MT, CE Domain 2: Reduction unit: DH, ER & R Domain 3: Palmitate release unit TE

FA Elongation:

Features	***Peculiarity***	***Energetics***
Enzyme: FA synthases or elongases Location: ER, mitochondria	Two carbon units can be added as far as 24 carbon atoms. Function to transform dietary essential FA to higher polyunsaturated FA (PUFA).	Per acetyl CoA → malonyl CoA = ***-1 ATP***

Unsaturated FA Synthesis

Features	***Peculiarity***	***Energetics***
Occurs in mitochondria and ER	Two carbon atoms can elongate them at a time.	A double bond is introduced directly into a saturated LCFA

Unsaturated FA are produced by terminal desaturases: Δ^9-, Δ^6-, Δ^5-, Δ^4- fatty acyl CoA desaturases exist	Cis double bond can be introduced not farther than C-9 from carboxylate end ***Function:*** Regulate physical properties of lipoproteins and membranes in regulation of metabolism in cells As precursors for eicosanoids	using NADH as cofactor: *- 3ATP per double bond introduced*

Branched FA Synthesis:

Features	***Peculiarity***	***Energetics***
BCFAs are synthesized in sebaceous and meibomian glands of human skin and are prominent components of many bacterial membranes.	***i. Branched -chain fatty acid synthesizing system: uses α-keto acids:*** (derived from transamination and decarboxylation of valine, leucine, and isoleucine) as primers ***ii. Branch-chain fatty acid synthase:*** utilizes *short-chain acyl-CoA esters* as primers	Requires malonyl CoA and NADPH per branch

Ketone Body Metabolism:

Features	***Peculiarity***	***Energetics***
Produced in liver as a consequence of prolonged metabolism of fat. Metabolized in other tissues as source of energy, cannot be utilized by (lacks enzymes: thiophorase for their utilization)	Preferred energy substrate for heart, skeletal muscle and kidney	*One acetoacetate* produces two acetyl CoA, each can generate about, 12 ATP *via* TCA of *i.e.*, total 24 ATPs activation of acetoacetate results in generation of one less ATP ***Net 23 ATP*** When β-hydroxy butyrate is oxidized, additional 3 ATP are formed by producing NADH

<table>
<tr><th colspan="3">Cholesterol Synthesis</th></tr>
<tr><th>Features</th><th>Peculiarity</th><th>Energetics</th></tr>
<tr><td>Ubiquitous, present in all membrane.

Occurs in cytosol and ER.

Occurs in liver, also in adrenal cortex, gonads and placenta.

Starts from acetyl CoA.

Requires NADPH and ATP.</td><td>Four stages of cholesterol synthesis:

1. Mevalonate synthesis: Acetyl CoA → HMG CoA by HMG CoA synthase. HMG CoA → mevalonate (C6) by HMG CoA reductase

2. Mevalonate is converted to isopentenyl diphosphate (isoprene, C5)

3. Six molecules of isoprene (C5) → squalene (C30)

4. Squalene undergoes cyclization with removal of three carbon atoms to yield cholesterol (C27)</td><td>18 moles of acetyl CoA, 36 moles of ATP and 16 moles of NADPH are required to synthesize one mole of cholesterol</td></tr>
</table>

<table>
<tr><th colspan="2">Cholesterol Breakdown</th></tr>
<tr><th>Features</th><th>Peculiarity</th></tr>
<tr><td>Cholesterol cannot be catabolized by mammalian cell.

Sterol ring of cholesterol cannot be degraded.

Cholesterol is either excreted in free form or in form of bile acid.</td><td>50% of cholesterol excreted as bile acids

Rest excreted as coprostanol and cholesterol

98-99% of bile acids reabsorbed: EHC (Enterohepatic Circulation) and rest excreted

First step: Degradation of cholesterol to bile acids by 7-α-hydroxylation

Hydroxylation at C_{12}, followed by reduction of C_3 carbonyl group

A series of oxidative steps in C_{20} to C_{27} side chain and oxidation to a carboxyl of alcohol group at C_{26} → cholic acid and chenodeoxy cholic acid</td></tr>
</table>

REGULATION OF PATHWAYS		
Pathway	***Enzyme***	***Regulators***
Fat Oxidation	-	*Most important regulator*: NAD/NADH$^+$H$^+$ ratio *Primary signals:* *Hormone sensitive lipase (HSL)*: (-) insulin (+) glucagon, epinephrine *Secondary signals:* Malonyl CoA inhibits carnitine acyl transferase
Fat Synthesis	ACC (Acetyl CoA carboxylase)	(+) Insulin (-) Glucagon, epinephrine; phosphorylation Palmitoyl CoA(-) Citrate (+)
Ketone Body Metabolism	ACC CPT1	*Low in fed state*: depression of FA oxidation *High in starvation*: increases FA oxidation *Inhibited* by acyl CoA in starvation→ decreasing malonyl CoA Decreased insulin/glucagon ratio inhibits ACC *ACC & decreased malonyl CoA levels→ activates CPT-1 allowing fatty acyl CoA to enter β-oxidation* *Fed state*: Malonyl CoA inhibits CPT-1 *Fasting*: Malonyl CoA decreases, CPT-1 inhibition Released
Cholesterol Synthesis	HMG CoA reductase	(+) insulin, thyroxine (-) glucagon, cholesterol

Cholesterol Breakdown	7 α-hydroxylase	(-) bile salts, vitamin C (+) cholesterol

FAT OXIDATION: Beta Oxidation	***Facts***
Steps ***Activation*** In cytosol, FA are activated to form fatty acid CoA by enzyme fatty acyl synthase. R–CH_2–CH_2– COO+ CoA+ ATP ↓ (ACS Acyl CoA synthase) R-CH_2 –CH_2 – C (=O)-S CoA + AMP+ PPi	It is located in mitochondria of all tissues and strictly aerobic. • Four repeat reactions for every two carbons, these reactions resemble those of TCA cycle. • Connection: fatty acid oxidation → acetyl CoA, $FADH_2$ and $NADH_2$ (from each round of oxidation). • Cleaves two carbons at a time from carboxyl end.

Carnitine Shuttle	*Enzymes of Carnitine Shuttle:*
It transfers fatty acids from cytosol to mitochondria. *Carnitine* Carnitine is β-hydroxy trimethyl ammonium butyrate.	*CPT I* [Carnitine palmitoyl transferase]: membrane – bound enzyme *CPT II*: resides on inner surface of inner mitochondrial membrane releases fatty acyl CoA and carnitine into matrix. *Translocase* enzyme: Transports acyl carnitine to CPT II. Free carnitine is transported in the opposite direction towards intermembraneous space.

Fatty acyl CoA released in mitochondrial matrix undergoes β-oxidation *via* cycle of four back-bone reaction (to release one acetyl CoA per cycle) sequence namely:

1. Dehydrogenation or oxidation: *FAD* takes H to create new double bond between C2 and C3.	***2. Hydration:*** Adds water across double bond.	***3. Oxidation:*** NAD takes H to form new carbonyl bond.	***4. Carbon – carbon cleavage/ Thiolase step:*** New CoA is added to cleave off

<table>
<tr><td colspan="2">Energetics:</td><td></td><td>acetyl CoA (i.e., loss of 2 carbons).</td></tr>
<tr><td colspan="3">Each cycle of β-oxidation reduces length of fatty acid chain by 2 carbons to produce:</td><td>= 1 acetyl CoA (12 ATP), 1 $FADH_2$ (2 ATP) and 1 NADH (3 ATP).</td></tr>
<tr><td colspan="3">Each 2-carbon unit of fatty acid finally produces 17 ATPs!

A 16- carbon fatty acid makes 8 acetyl CoA i.e., 12 x 8 = 6

7 cleavage points release 5 ATP (1NAD & $1FADH_2$ yield 5ATP) = 7x5 = 35 ATP</td><td>Total ATP produced= 96 + 35 = 131

Two ATP are used = 131 -2

Net ATP produced= 129</td></tr>
<tr><td colspan="4">Oxidation of odd chain FA yields propionyl CoA → succinyl CoA → TCA cycle.</td></tr>
<tr><td colspan="4">Minor FA Oxidation Pathways</td></tr>
<tr><td colspan="2">Alpha Oxidation

Only one carbon is removed at carboxyl end.

It is not linked to high energy production.

It requires monooxygenases, ascorbic acid and Fe^{+2} to produce α-hydroxy acid.</td><td colspan="2">Omega Oxidation

Omega (last) carbon at carboxyl end is oxidized to corresponding alcohol to form dicarboxylic acid.

It occurs in endoplasmic reticulum and requires cyt P_{450}, oxygen and NADPH to form dicarboxylic acid.</td></tr>
<tr><td colspan="3">FAT SYNTHESIS:</td><td rowspan="2">Reactions:

Reactions of fatty acid synthesis all occur on one enzyme: fatty acid synthase (FAS) to yield free fatty acid (FFA). FAS enzyme is multienzyme complex: dimer of two identical polymonomers.</td></tr>
<tr><td>Location: Cytosol

Organs: Liver, kidney, mammary gland, adipose tissue, brain

Substrate: Acetyl CoA</td><td colspan="2">Cofactors: NABPH, ATP, Biotin, Mn^{+2}, HCO_3

End product: palmitate</td></tr>
</table>

1st Step:	***FAS (Fatty Acid Synthase)***
Formation of malonyl CoA is initial and rate-controlling step of FA synthesis by enzyme acetyl CoA carboxylase (ACC).	Multi enzyme complex, dimer with six enzyme activities: ATAcetyl transacylase MTMalonyl transacylase CEcondensing transacylase TEThioesterase DH dehydratase ER Enoyl reductase KRketo acyl reductase
CH_3COS CoA + ATP + HO_3 Acetyl CoA ACC ↓ ATP, CO_2 biotin → ADP 3 $OCCH_2COS$ CoA Malonyl CoA	
Acetyl CoA Carboxylase (ACC): ACC step is committed step, key regulatory step, and irreversible step. It is regulated by palmitoyl CoA (inhibits), citrate (activate), acyl CoA (negative feedback inhibition), insulin (activates).	Each single polypeptide monomer has with 3 distinct domains: Domain 1: Substrate entry and condensation unit AT, MT, CE Domain 2: Reduction unit: DH, ER & R Domain 3: Palmitate release unit TE.
KETONE BODIES	Produced in liver as a consequence of prolonged metabolism of fat. Metabolized in other tissues as source of energy. Cannot be utilized by liver as liver lacks enzymes (thiophorase) for their utilization.
Ketone bodies include: Acetone: $CH_3-C(=O)-CH_3$	

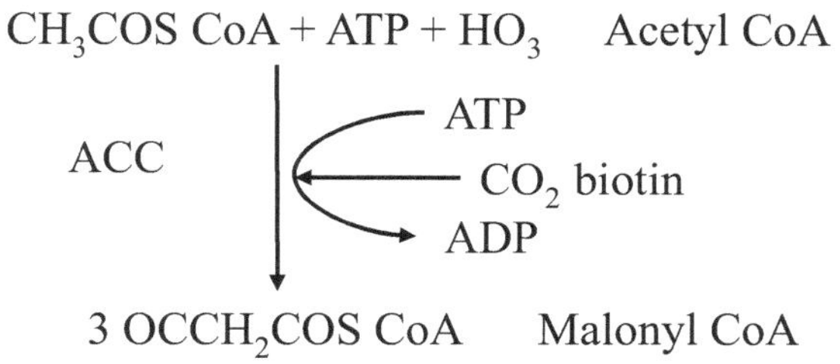

Acetoacetate: $CH_3 - C - CH_2 - COO$ (with $\| O$ on the C)

Beta hydroxy butyrate: $CH3 - CH - CH2 - COO$ (with $| OH$ on the CH)

Metabolism

Ketone bodies are generated by liver and utilized by muscle and brain.

Ketone Body Utilization

Utilized in other tissues (but not liver):

Acetoacetate → acetoacetyl CoA (*via thiophorase*)

Acetoacetyl CoA → two acetyl CoA (thiolase)

Beta Hydroxybutyrate → acetoacetate + ATP (dehydrogenase).

Acetyl CoA then enters TCA cycle and gets oxidized to molecules of CO_2.

Ketone Body Synthesis:

Occurs during a state of high rate of fat oxidation in liver e.g. starvation, prolonged exercise.

Formed in liver mitochondria by condensation of three acetyl CoA units by thiolase enzyme.

Then flows to extrahepatic tissues after formation.

Acetoacetate spontaneously decarboxylated non-enzymatically to acetone.

Steps:

Two acetyl CoA units condensed → *acetoacetyl CoA*

Acetoacetyl CoA + acetyl CoA --- *HMG CoA synthase*
→*HMG Co (hydroxymethylglutaryl CoA)*

HMG CoA *split* by *HMG CoA lyase*
→ acetyl CoA + acetoacetate.

Acetoacetate ---- *dehydrogenase enzyme*
→ Hydroxybutyrate

CHOLESTEROL SYNTHESIS

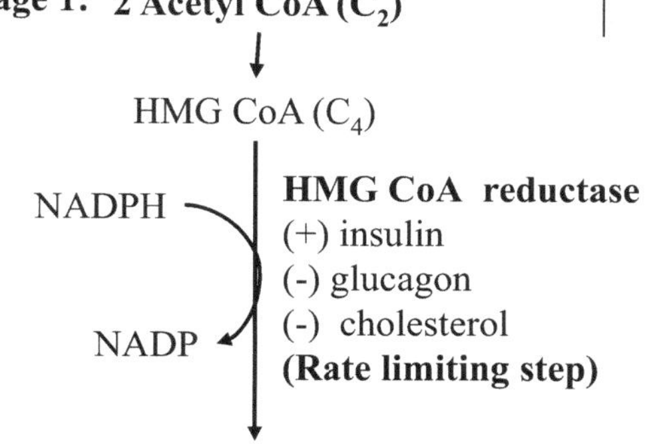

Stage 2: Mevalonate

3ATP

3ADP

CO_2

Isoprenoid (C_5)

NADPH

NADP

Active Isoprenoid (C_5)

Squalene (C_{30})

Stage 3: Squalene C_{30}

NADPH

O_2

NADP

Cyclization

Squalene epoxide cyclase

Lanosterol (C_{30})
(4-ring)

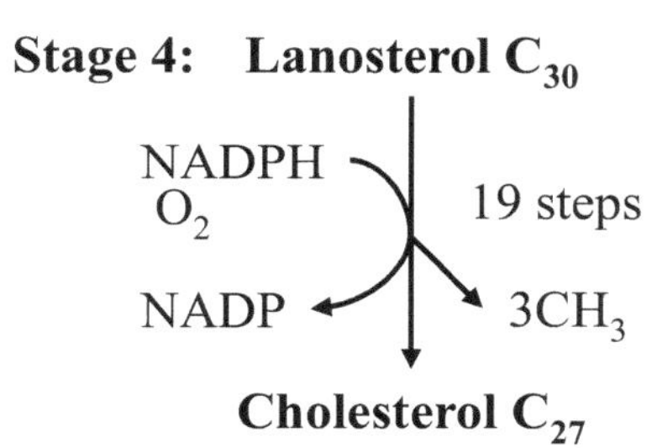

Conversion of Cholesterol to Bile Acids:

The first step in degradation of cholesterol to bile acids to 7-α-hydroxylation reaction that requires NADPH and molecular O_2.

7 α hydroxylase:

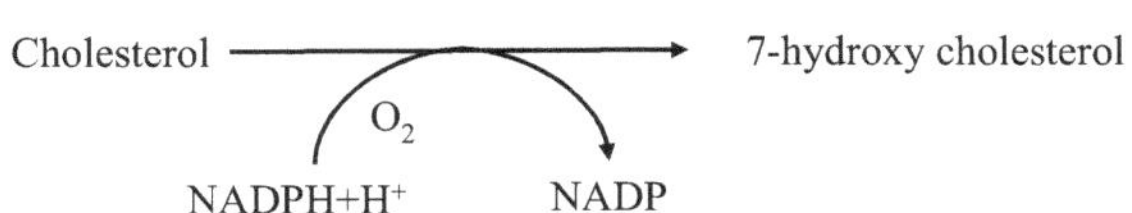

Functions of Bile Salts

Emulsification of fats due to detergent activity.

Aid in absorption of fat-soluble vitamins.

7α-hydroxylase:

First committed and rate-limiting step.
Microsomal (monooxygenase) enzyme.
Repressed by bile salts.
Induced by cholesterol.

Accelerate action of pancreatic lipase. Stimulate intestinal motility. Keep cholesterol in solution (as micelles).	Vitamin C interferes with this step.

<u>*Steps:*</u>

- Hydroxylation at C_{12} for precursor of cholic acid.
- Reduction of C_3 carbonyl group.
- A series of oxidative steps occur in C_{20} to C_{27} side chain and oxidation to a carboxyl of alcohol group at C_{26}, thus forming ***cholic acid*** (3, 7, 12 - α-triihydroxycholanic acid) and ***chenodeoxycholic acid*** (3, 7-α-dihydroxycholanic acid).
- They are conjugated with taurine or glycine to become more water soluble and get excreted in bile to EHC.

LIPOPROTEIN METABOLISM & ASSOCIATED DISORDERS

Introduction	**Lipoproteins-structure**
Major function of lipoproteins is transport of energy- rich triglyceride from intestine and liver to the sites of storage and utilization.	• Nonpolar lipid core of triacylglycerol and/or cholesterol ester surrounded by amphipathic lipids (phospholipids and cholesterol) and proteins. • Protein moiety known as apolipoprotein or apoprotein. • Apoprotein constitute 70% of HDL and 1% of chylomicrons. • Apolipoprotein can be integral (Apo B) or free loosely bound (apo C).

Lipoprotein	***Function***
Chylomicron	Deliver FA as part of TAG from dietary fat to muscle, adipose tissue.
Chylomicron remnant	Deliver dietary cholesterol to liver.
VLDL	Deliver FA attached to TAG, derived from liver synthesis.

LDL	Formed from VLDL delivers cholesterol, derived from liver synthesis to various tissues.	
HDL	Collects (scavenges) cholesterol from non-hepatic tissues and delivers to liver.	
Lp in Human Disease		
Abetalipoproteinemia	***Rare Genetic Disorder*** Absence of all apo B-containing lipoproteins (chylomicrons, VLDL and LDL). Apo B is essential for chylomicron VLDL formation. Defect in TAG transfer protein prevents loading of apo B with lipid and lipoprotein containing Apo B not formed. Lipid droplets accumulate in liver and intestine.	***Features*** Fat malabsorption Accumulation of lipid droplets in enterocytes Spiny shaped RBCs Neurological disease (ataxia and retardation)
LCAT deficiency	Due to inability to convert cholesterol associated with HDL to CE. Normally, CE are transferred to other LP which would then be taken up by receptors in liver. Continued removal of cholesterol from periphery requires LDL.	*Result:* defects in kidney, RBC and cornea of eye.
Familial hypercholesterolemia (FH)	Autosomal dominant disorder Characterized by decreased number of LDL receptors Mutation in gene coding for apo BIE (LDL receptor)	*Features:* Patient has high plasma cholesterol and LDL cholesterol and carries a high risk of premature coronary disease

Dyslipidemia			
Type	***Inheritance***	***Defect***	***Outcome***
Familial hyper cholesterolemia	AD	LDL-receptor defect	Increase cardiovascular risk, raised cholesterol levels
Familial combined dyslipidemia	AD	Overproduction of apo B 100	Increase cardiovascular risk, raised TAG level
Dysbetalipoproteinemia	AR	Defective remnant binding to LDL-receptor	Increased cholesterol and TAG levels, increased cardiovascular risk

Hyperlipidemia		
Type	***Lipoprotein Increased***	***Defect***
I	CMN	Familial lipoprotein lipase deficiency (chylomicron syndrome)
IIa	LDL (familial)	Defect in LDL ®: mutation in ®
IIb	LDL, VLDL	Defective apo B gene
III	IDL familial combined hyperlipidemia	u/k* accumulation of remnant-like particle, apo B. also called dysbetalipidemia
IV	VLDL (familial hypertriglyceridemia)	u/k, Chylomicronemia, lipoprotein lipase deficiency), high triglyceride levels
V	CMN, VLDL	u/k*, overproduction of VLDL or in VLDL conversion to LDL

ATHEROSCLEROSIS	A general term describing any hardening or loss of elasticity of medium or larger arteries and refers to formation of atheromatous plaque.

<table>
<tr><td>Risk Factors

Diabetes or impaired glucose tolerance

Dyslipoproteinemia:

-High LDL

-Low HDL

-LDL / HDL ratio > 3:1

Smoking

High blood pressure

Advancing age

Familial hypercholesterolemia</td><td colspan="2">Features

A multifactorial complex disease that is triggered and maintained by a low-level chronic inflammation of the arterial wall.

Key players in development of hyperlipidemia and atherogenesis: LDL and HDL cholesterol

There is development of atheromatous plaque.

It is characterized by accumulation of fatty substances called plaque.

Plaque is made up of excess fat, collagen and elastin.

As plaque grows, artery wall thickening occurs and narrowing of lumen occurs.

Plaque can rupture forming a thrombus, which can slow or stop blood flow causing infarction.

Most commonly plaque formation occurs in coronary artery or arteries of brain.</td></tr>
<tr><td>Causes of Endothelial Dysfunction</td><td>Modified LDL

Homocysteine

Free radicals

Infection</td><td>Hypertension

Once endothelium is irritated, an inflammatory response follows.

Ist phase of inflammatory response is increased permeability to lipoproteins and other plasma constituents.</td></tr>
</table>

<table>
<tr><td rowspan="2">Diagnostic uses of Lipid and Lipoprotein Values</td><td>First Line Screening Test:</td></tr>
<tr><td><u>Step 1</u>: Total cholesterol</td></tr>
</table>

Measurement of Plasma Lipids Indicated in ***Individuals:***	***Step 2:*** Fasting lipid profile
• CHD, cerebrovascular and peripheral vascular disease. • Family history of premature CAD. • Other risk factors for CHD: HT. • DM. • Patients with clinical features of hyperlipidemia. • Plasma of patient being lipemic.	Total cholesterol Total triglycerides HDL-C calculate: Cholesterol/HDL-C ratio LDL-C
	Step 3: VDL-C LDL-C HDL-C

EICOSANOIDS	
A complex family of bioactive lipid messengers generated mainly from arachidonic acid. Two main classes: (i). Prostanoids: PG, TX, PGI_2. (ii). Linear eicosanoids: LT, LX, HETE.	Arachidonic acid (AA) is usually derived from C-2 position of PL in plasma membrane by action of PLA_2. AA is then converted to PG_2, TX_2, LT_4, and LX_4. PG are designated as PGA, PGD, PGE or PGF based on functional groups on cyclopentane ring.

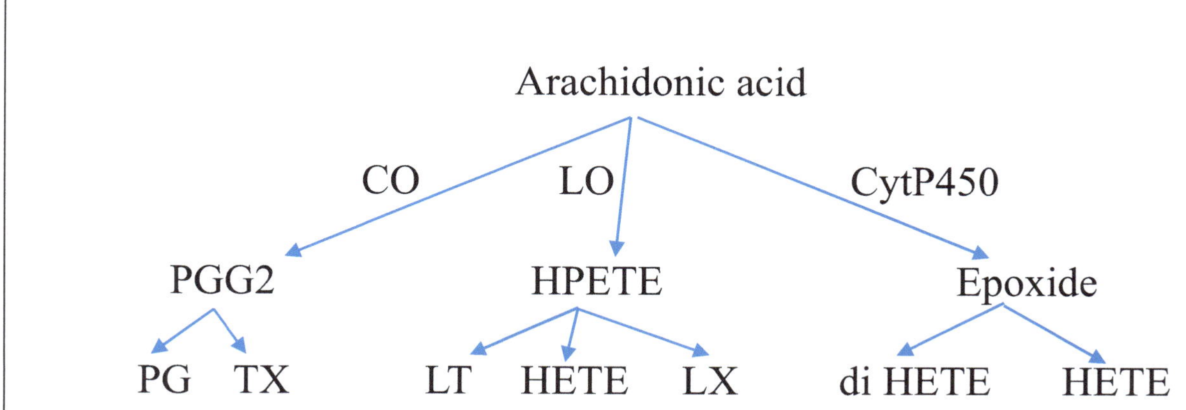

Pathway of Eicosanoid Metabolism

Synthesis of Eicosanoids	By action of phospholipase A_2 (PLA_2), arachidonic acid and lysophosphatidic acid are released from membrane phospholipids from *sin* -2 position. The key enzyme of PG synthesis is PG G/H synthase (PGS).

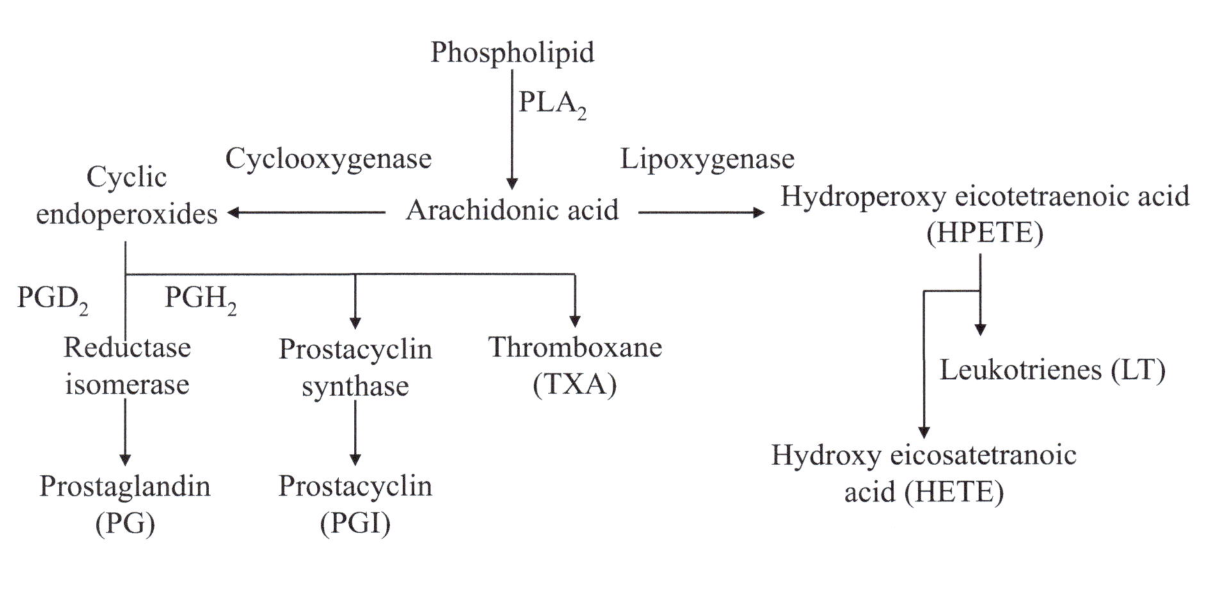

Prostanoid	**Action**
PGD_2	Found in mast cells and nervous system Vasodilation Inhibit platelet aggregation Relaxation of GI muscle Uterine relaxation
$PGF_{2\alpha}$	Myometrial contraction Luteolysis (cattle) Bronchoconstriction (dog)
PGI_2	Synthesized in vascular and gastric tissue. Vasodilation. Pulmonary vasodilator. Inhibit platelet aggregation: antiaggregator. Rennin release and natriuresis: *via* effects on tubular reabsorption of Na^+.
TXA_2	Vasoconstriction Stimulate platelet aggregation Bronchoconstriction
PGE_2	Produced in most cells and increase cAMP levels. - EP_1 receptors: bronchial and GI smooth muscle contraction. - EP_2 receptors: bronchodilation, vasodilation, (GI) smooth muscle relaxation (+) intestinal fluid secretion. - EP_3 receptors: smooth muscle contraction. inhibit gastric acid secretion increase gastric mucus secretion

	stimulate uterus contraction
Therapeutic Uses of Prostaglandins	
Prostaglandin E1	Misoprostol (oral, Intravaginal) Gemeprostol Alprostadil (IV, Intraurethral, Intracavernosal route)
Prostagladin I2	Epoprostenol (IV)
Prostaglandin F2 α	Latanoprost (Opthalmic) Carboprost (IM)
Prostaglandin E2	Dinoprostone (Intravaginal)

QUESTIONS

1. Short notes :

   a. Pancreatic lipase
   b. Micelle
   c. Fat malabsorption
   d. Chylomicrons
   e. Fatty liver
   f. Ketosis
   g. Refsum's disease
   h. Carnitine
   i. Reverse cholesterol transport
   j. PUFA

2. What are the different type of lipases?
3. Explain how fats help in absorption of fat?
4. Differentiate between fatty acid synthesis and fatty acid oxidation.
5. What are the sources of NADPH required for fatty acid synthesis?
6. Make a list of multiunit enzyme complexes.
7. Describe regulation of fatty acid synthesis.
8. What is the importance of carnitine in fat metabolism?
9. What are different types of fatty acid oxidations?
10. Name fatty acid oxidation defects.
11. What is the importance of omega oxidation of fatty acids?
12. Make a list of isoprenoid compounds.
13. What are the fates of cholesterol?
14. Name primary and secondary bile acids.
15. How do PUFA and MUFA affect cholesterol levels?
16. Name the lipid lowering drugs.
17. What are different lipoproteins and what are their functions in our body?
18. What is reverse cholesterol transport?
19. How does oxidized LDL contributes to atherosclerosis?
20. What is fatty liver?
21. Give the importance of essential fatty acids and disorder associated with their deficiency.

22. Describe beta oxidation of fatty acids. How many ATPs are released on complete oxidation of palmitic acid?

23. Discuss in brief:

    a. Role of bile salts in digestion of fats

    b. Sources and fate of acetyl CoA

    c. Methyl malonic aciduria

    d. Ketone body metabolism

    e. Lipoproteins

    f. Atherosclerosis

    g. Sources and uses of cholesterol.

    h. HMG CoA reductase.

24. Discuss denovo synthesis of fats.

25. Discuss cholesterol synthesis.

BIBLIOGRAPHY

Denise R Ferrier. Lippincott illustrated reviews: biochemistry. 7th Edition. Philadelphia Wolters Kluwer; 2017

Donald Voet, Judith G Voet, Charlotte W Pratt. Fundamentals of Biochemistry. 5th Edition. New York: Wiley; 2016.

Geoffrey L Zubay, Dennis E Vance. Principles of biochemistry. Dubuque, Iowa: William C. Brown; 1995.

Jeremy M Berg, Gregory J Jr Gatto, Lubert Stryer, John L Tymoczko. Biochemistry. 9th Edition. New York: Macmillan International Higher Education: WH Freeman; 2019.

Lehninger A, Nelson D, Cox M. Lehninger principles of biochemistry. New York: Worth Publishers; 2000.

Michael A Lieberman, Rick E Ricer. Biochemistry, molecular biology, and genetics. 7th Edition. Philadelphia, Pa Wolters Kluwer; 2020.

Victor W Rodwell, David A Bender, Kathleen M Botham, Peter J Kennelly, P Anthony Weil. Harper's illustrated biochemistry. 31st edition. New York: Mcgraw-Hill Education; 2018.

Mind Maps in Biochemistry, 2021, 120-125

CHAPTER 7

Chemistry of Proteins

LEARNING OBJECTIVES:	Keywords:
• Appraise the protein structure and function. • Summarize the properties and classification of amino acids. • Illustrate the classification of proteins. • Identify the diseases related to structural anomalies in proteins.	Amino acids – optical activity, Amphoteric nature, Alpha-helix, Beta-sheet, Diseases related with structural anomalies in proteins, Fibrous and globular proteins, Glycogenic, Ketogenic amino acids, Non- polar, essential, Non-essential, Polar, Primary, Protein misfolding, Proteins, Secondary, Tertiary and quaternary structure of proteins.

PROTEINS: STRUCTURE AND FUNCTION

Functions of Proteins	**Amino Acids**	**H** | **H_2N – C – COOH** | **R**
Structural: Make up cytoskeleton - Component of collagen, elastin, keratin Enzyme catalysis Transport Storage	Functional units of proteins Composed of: Two functional groups: Amino ($-NH_2$) Carboxyl (-COOH)	Fig. (**7.1**) structure of amino acid: - NH_2 = amino group (basic) - COOH = carboxyl group (acidic) - R = side chain

Simmi Kharb

Hormone Blood coagulation Immunity Control of gene expression	Hydrogen atom (-H) Distinctive side-chain (-R), attached to a central carbon termed *α-carbon*. Except for glycine, all acids contain at least one asymmetric carbon atom (α-carbon atom)	Of these 20 amino acids, only proline is imino acid (-NH-), and not an α-amino acid.

<table>
<tr><td colspan="2">PROPERTIES OF AMINO ACIDS</td></tr>
<tr><td>Optical Activity

Amino acids exhibit optical isomerism: enantiomers

Two amino acid configurations: D- (dextro or right) and L- (levo or left).

Only L - α-amino acids occur in proteins</td><td>Amphoteric

Exist as Zwitter ions in solutions at neutral pH (both positive and negative charges)

α-carboxyl group is negatively charged, and α-amino group is positively charged</td></tr>
</table>

<table>
<tr><td colspan="4">CLASSIFICATION OF AMINO ACIDS</td></tr>
<tr><td>Based on Structure:</td><td colspan="2">On the basis of the side chain (-R) attached to α-carbon atom:-</td><td rowspan="5">Based on their Metabolic Fate

Glycogenic

(Ala, Arg, Asp, Cys, Glu, Gly, His, Met, Pro, Ser, Thr, Val)

Ketogenic

(Leu)

Both

(Ile, Lys, Phe, Trp, Tyr)</td></tr>
<tr><td colspan="2">Polar (Hydrophilic): Glycine, Serine, Threonine, Cysteine, Arginine, Histidine, Lysine, Aspartate, Asparagine, Glutamate, Glutamine</td><td>Non polar (Hydrophilic): Alanine, Leucine, Isoleucine, Valine, Methionine, Proline, Tyrosine, Phenylalanine, Tryptophan</td></tr>
<tr><td colspan="3">Based on Nutritional Requirement</td></tr>
<tr><td>Essential Amino Acid: Cannot be synthesized by the body and must be supplied in diet - Methionine, Arginine, Threonine, Tryptophan, Valine, Isoleucine, Leucine,</td><td colspan="2">Non-essential Amino Acid: Can be synthesized by the body to meet its demands - Glycine, Alanine, Serine, Cysteine, Aspartate, Asparagine, Glutamate, Glutamine, Tyrosine, Proline</td></tr>
</table>

Phenylalanine, Lysine, Histidine		

STRUCTURAL ORGANIZATION OF PROTEINS

Two Categories of Proteins

Fibrous proteins- structural proteins, are more filamentous or elongated *E.g.*, connective tissue, tender, muscle fibers	*Globular protein* – compactly folded and coated and are water soluble. They act as transporters.

Key levels of proteins are primary, secondary and tertiary structures and quaternary structure

Primary Structure	***Secondary Structure***	***Tertiary Structure***	***Quaternary Structure***
Synthesized on ribosome as a linear sequence of amino acids in a polypeptide chain. Determined by the sequence of amino acids. Serves as a foundation for higher levels of protein structure.	Short segments of polypeptide chain folds to form secondary structure. Achieved through weak bonds *e.g.*, hydrogen bonding. Types: - α helix - β sheets - β turn	Entire 3- dimensional conformation of the polypeptide Its structure indicates how secondary structure features assemble and relate to each other in 3-dimensional space.	The structure is formed as a result of interactions between two or more polypeptide chains that give rise to a specific geometry and the aggregate formed is called *oligomer e.g.*, hemoglobin.

α HELIX	***β - PLEATED SHEET (or Conformation)***
Cylindrical in shape Formed by coiling of polypeptide chain on itself at every fourth peptide linkage	The polypeptide chain is fully stretched out and can fold on itself with its segments packed together.

<table>
<tr><td colspan="2">Formed by H-binding

Can be right-handed helix

Provides some rigidly to that portion of the molecule.

E.g., collagen is composed of triple helix.</td><td colspan="2">β- TURN OR RANDOM COIL

It is formed by folding of the polypeptide chain

Has little stability due to the presence of certain amino acid whose side chain interferes with one another.</td></tr>
<tr><td colspan="4">STRUCTURE- FUNCTION RELATION OF PROTEINS</td></tr>
<tr><td colspan="4">The primary structure of a protein determines its spatial structures and biological functions.

Proteins having similar amino acid sequences have functional similarity.

Any change in key AAs in a protein can cause loss of its biological functions.

Also, particular spatial structure of a protein is correlated with its specific biological functions.

Examples:</td></tr>
<tr><td>Denatured Proteins:

Their primary structure is retained, but, they have no biological function.</td><td>Allosteric Change of Hemoglobin by O₂</td><td colspan="2">Protein Misfolding and Degenerative Diseases: An error in protein conformation can lead to disease:

- Abnormal conformational transition from alpha helix to beta sheet: characteristic of amyloid deposits.
- Protein misfolding: occurs in Huntington's disease.</td></tr>
<tr><td colspan="3">Hb structure and sickle cell disease: structural abnormality in β globin chain.

<u>Single point mutation</u>: glutamic acid at sixth position in β chain is replaced by valine</td><td>Qualitative hemoglobinopathy: Defect in production of α or β-chain

- α-Chain deficiency: α-Thalassemia
- β- Chain deficiency: β-Thalassemia</td></tr>
</table>

QUESTIONS

1. Discuss classification of amino acid with suitable examples

2. What are essential amino acids and how they differ from non-essential amino acids?

3. What are zwitterions?

4. Discuss briefly:

   i. Essential amino acids.

   ii. Amphoteric nature of amino acid.

   iii. Iso-electric pH (pI)

   iv. Protein color reactions

5. Short notes:

   i. Primary structure of insulin

   ii. Biological importance of proteins

   iii. Precipitation of proteins

   iv. Denaturation of proteins

   v. Peptide bond

6. Classify proteins with suitable examples.

7. Give an account of structure of proteins.

BIBLIOGRAPHY

Denise R Ferrier. Lippincott illustrated reviews: biochemistry. 7th Edition. Philadelphia Wolters Kluwer; 2017

Donald Voet, Judith G Voet, Charlotte W Pratt. Fundamentals of Biochemistry. 5th Edition. New York: Wiley; 2016.

Geoffrey L Zubay, Dennis E Vance. Principles of biochemistry. Dubuque, Iowa: William C. Brown; 1995.

Jeremy M Berg, Gregory J Jr Gatto, Lubert Stryer, John L Tymoczko. Biochemistry. 9th Edition. New York: Macmillan International Higher Education: WH Freeman; 2019.

Keith Wilson, John M Walker. Principles and techniques of biochemistry and molecular biology. 7th edition. Cambridge: Cambridge University Press; 2017.

Lehninger A, Nelson D, Cox M. Lehninger principles of biochemistry. New York: Worth Publishers; 2000.

Michael A Lieberman, Rick E Ricer. Biochemistry, molecular biology, and genetics. 7th Edition. Philadelphia, Pa Wolters Kluwer; 2020.

Victor W Rodwell, David A Bender, Kathleen M Botham, Peter J Kennelly, P Anthony Weil. Harper's illustrated biochemistry. 31st edition. New York: Mcgraw-Hill Education; 2018.

CHAPTER 8

Metabolism of Proteins

LEARNING OBJECTIVES:	Keywords:
• Describe the assimilation of dietary proteins. • Illustrate the role and regulation of enzymes of protein metabolism. • Identify the disorders of protein metabolism and interpret clinical correlations. • Explain the metabolic fate of amino acids.	Amino acid metabolism, Digestion of proteins, Inherited metabolic disorders and other diseases of protein metabolism, Proteolytic enzyme, Zymogen.

DIGESTION AND ABSORPTION OF DIETARY PROTEINS

Sources of Amino Acids	Fact File
Dietary/ Exogenous Proteins: Include animal products (meat, poultry, fish and dairy products) and plant products (grains, legumes and vegetables). ***Endogenous Breakdown Protein:*** Include a breakdown of defective and unneeded cellular proteins. They can be desquamated mucosal cells, digestive enzymes and glycoproteins derived from secretions of salivary glands, stomach, intestine biliary tract and pancreas. Proteins of endogenous origin are digested and absorbed more slowly than that of exogenous origin.	***Dietary Protein Intake***: 70-100 g/day 50-60% of animal origin 20-30 g in vegetarians ***Proteins Secreted into Intestine***: 30 g of desquamated cells and 1~2 g plasma proteins, enzymes and mucoprotein Fecal excretion of protein: ~10g/day
Digestion of Proteins	The dietary proteins are cleaved by the action of proteolytic enzymes produced by stomach, pancreas and small intestine

Simmi Kharb

Enzymes for digestion of protein:	Pepsin, trypsin, chymotrypsin, elastase, carboxy peptides, amino peptides.
Digestion of protein begins in the stomach with the action of pepsin:	Produced as inactive zymogen pepsinogen by chief cells Activated to pepsin by H^+ and pepsin itself (auto catalysis) Optimum pH: 2.0
In the small intestine, protein digestion begins by enzymes:	Serine protease Trypsin Chymotrypsin Elastase Carboxy peptidase

Activation of Trypsinogen:	***Trypsin:***
Trypsinogen is activated by enterokinase secreted from intestinal brush border in response to secretin and CCK. Once trypsin is formed, it attacks additional molecules of trypsinogen and other zymogens: Chymotrypsinogen, proelastase, procarboxy peptidase to yield active proteolytic enzymes.	Endopeptidase Secreted by intestinal mucosa Hydrolysis of the lysine peptide bond in zymogen: - Releases a small peptide from trypsinogen - Allows molecule to unfold as active trypsin

Enzymes for Digestion of Proteins & Activation of Zymogen form of Proteolytic Enzyme and their Actions:

Zymogen	***Activators***	***Active form***	***Bond specificity***	***Site of action***
Pepsinogen	HCL, pepsin	pepsin	Most amino acid	Stomach
Trypsinogen	Enteropeptidase Trypsin	Trypsin	Basic amino acid	Intestine
Chymotrypsinogen	Trypsin	Chymotrypsin	Aromatic amino acid	Intestine

Proelastase	Trypsin	Elastase	Broad specificity	Intestine
Pro carboxy peptidase	Trypsin	Carboxy-peptidase A, B	Carboxyterminal (exopeptidase) aromatic, neutral amino acid, basic amino acids	Intestine

Digestion in Small Intestine Brush Border	Several peptidases are produced by the brush border of the small intestine and they complete the digestive process.
Enzymes of Small Intestine Brush Border:	Endopeptidase, Exopeptidase, Aminopeptidase, Tripeptidase, Dipeptidase.
Enzyme	***Mechanism of Action***
Carboxy peptidase	Exopeptidase attack carboxy-terminal peptide
Amino peptidase	Exopeptidase attack peptide bond next to the amino terminal.
Tripeptidase	Digest tripeptides to yield dipeptide and free amino acid.
Dipeptidase	Digest dipeptides to free amino acids.
End Products of Protein Digestion	Peptides, dipeptides, tripeptides, and free amino acids

Fate:	***Absorption of Amino Acids***	***Amino acid transporters present at brush–border:***
They get absorbed across the brush border of intestinal mucosal cells. Free amino acids and dipeptides can enter enterocyte by carrier-mediated membrane transport to finally enter the bloodstream.	L-isomer of amino acids is actively absorbed by active transport.	**1. *Na – dependent symporters*** **2. *H dependent symporters*:** H-dependent transport is 'E' dependent and may require pyridoxal phosphate.

Amino Acid Symporters:	
Symporter	***Amino Acids***
Neutral amino acid	Ser, Thr, Ala
Neutral amino acid with an aromatic amino acid side chain	Phe, Thr, Met, Val, Leu, Ile
Imino acid	Proline, Hydroxyproline
Basic amino acid	Lys, Arg, Cys
Acid amino acid	Asp, Glu
β amino acid	β-alanine, taurine

Gamma Glutamyl Cycle	
Fact File: Located in the cell membrane. Group transfer transport system for an amino acid. Utilizes 3 moles of ATP to transport a single amino acid. Functions in kidney mainly, also present in RBCs, neurons. Operative for the transport of cysteine and glutamine. Important for the metabolism of GSH (glutathione)	***Steps:*** *1.* Enzyme glutamyl transpeptidase is located in membrane and shuttles GSH to the cell surface to interact with an amino acid. *2.* – *a.* Glutamyl amino acid is transported into the cell and hydrolyzes to liberate amino acid and oxyproline *b.* Cysteinyl is cleaved to its component amino acid. *3.* To regenerate GSH, glutamate is reformed from 5- oxoproline and 2 ATP are required in this reaction.

Applied Aspects: Absorption	Absorption of undigested polypeptides may cause antigenic reactions.		
Diseases / Disorders of Protein Absorption			
Disease	***Defect***	***Inheritance***	***Mechanism***
Nontropical Sprue	Gluten absorption.	-	-
Celiac Disease	Protein and amino acid absorption.	-	-
Hartnup Disease	Neutral amino acid carrier defect.	Autosomal recessive	Essential amino acids not absorbed in the intestine and other neutral amino acids not absorbed in kidneys *Pellagra* (due to deficiency of niacin, tryptophan)
Cystinuria	Defect in transport of basic amino acid cysteine both in intestine and kidney.	Autosomal recessive	Cysteine concentrates in urine and forms *kidney stones* (cysteine crystal deposition)
Glycinuria	Defect in transport of glycine and imino acid proline and hydroxy proline.	Autosomal recessive	Increased excretion of amino acid No clinical abnormalities

PROTEIN METABOLISM

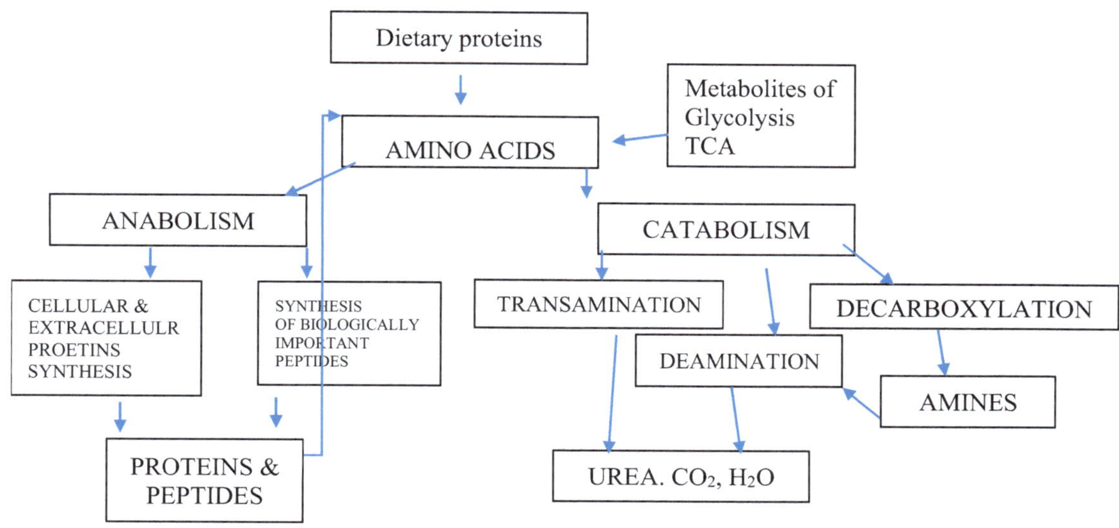

Fig. (8.1). General pathway of amino acid metabolism.

AMINO ACID CATABOLISM: CARBON SKELETONS (Figs. 8.1 and 8.2)

Fact File	**Amino Acid Carbon Skeletons**
Virtually all carbon skeletons can be converted into intermediates of Glycolytic pathway, TCA cycle, lipid metabolism. ***1st STEP*** Transfer of α amino group by transamination to αKG or OAA to provides glutamate and aspartate: source for 'N' in urea cycle. Exceptions: Lysine, proline, hydroxyproline, threonine: do not undergo transamination.	Amino acids are deaminated to α keto acids that are fed into major metabolic pathways (*e.g.* TCA) Amino acids can be grouped as: Glucogenic Ketogenic Based on whether or not their carbon skeletons can be converted to glucose.

Glucogenic Amino Acid	*Ketogenic Amino Acid*
Carbon skeletons of glucogenic amino acid are degraded to pyruvate, 4C, 5C-intermediates of TCA cycle. These carbon skeletons serve as precursor for glucose *via* GNG.	Carbon skeletons of ketogenic amino acids are degraded to acetyl CoA, acetoacetate. They cannot be converted to glucose.

KETOGENIC

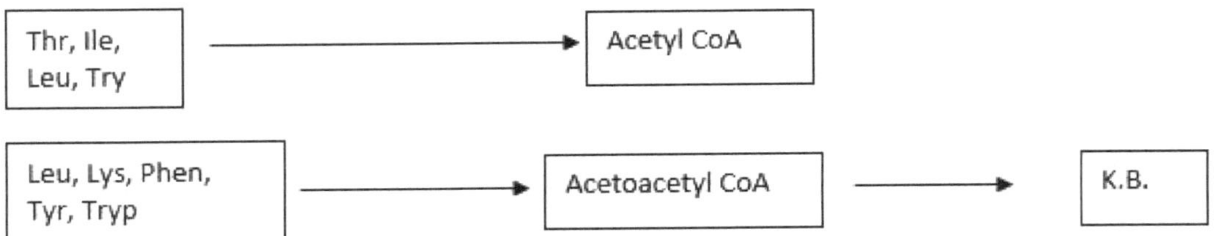

GLUCOGENIC

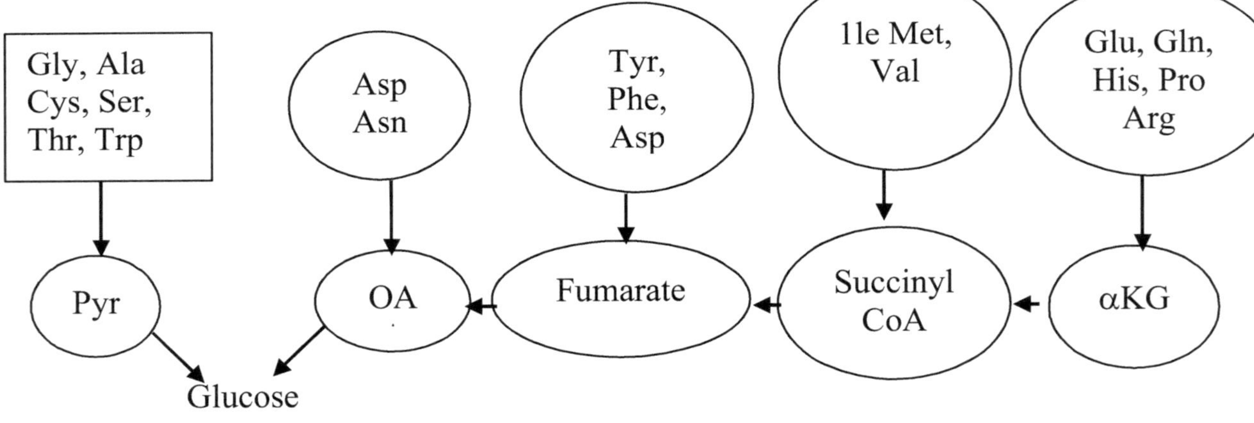

Fig. (8.2). Metabolic fate of carbon skeletons.

<table>
<tr><td colspan="3">Inherited Metabolic Disorders (IMD)</td></tr>
<tr><td>Amino Acid Catabolism

PKU (Phenylketonuria)

Alkaptonuria (Phe, Tyr)

Tyrosinemia (Phe, Tyr)</td><td>Methylmalonyluria (Met, Val, Ile)

Isovalterate academia (Ile, Leu, Val)

Histidinemia (His)

Hyperprolinemia (Pro)</td><td>IMD of amino acid results in:

• Accumulation of intermediate in blood or urine.

• Severity of disease depends on toxicity of accumulated metabolites

Treatment: Low protein diet</td></tr>
<tr><td colspan="2">GENERAL ASPECTS OF AMINO ACID BIOSYNTHESIS</td><td>COMMON DISORDERS ASSOCIATED WITH PROTEIN METABOLISM</td></tr>
<tr><td colspan="2">• All amino acids are synthesized from common metabolic intermediates

• Non-essential amino acids are synthesized by transamination of α-keto acids that are available as common intermediates.

• * α-keto acids of essential amino acid are not common intermediates (enzymes needed to them are lacking), transamination.</td><td>Hereditary Deficiency of Urea Cycle Enzymes

• Hereditary deficiency of any of the urea cycle enzymes leads to: hyperammonemia – elevated ammonia in blood

• Total lack of any of the enzyme of urea cycle is lethal

• Elevated ammonia is toxic, especially to brain.

• If elevated ammonia is not treated immediately after birth, severe mental retardation results.</td></tr>
<tr><td colspan="2">Mechanism for Toxicity of High Ammonia:</td><td>Treatment of Deficiency of Urea Cycle Enzyme</td></tr>
<tr><td colspan="2">1. High ammonia drive glutamine synthase:

$Glu + ATP + NH_3 \rightarrow Gln + ADP + Pi$</td><td>• Limit protein intake</td></tr>
</table>

This *depletes glutamate*, a neurotransmitter (excitatory) and precursor for synthesis of (inhibitory) neurotransmitter GABA. 2. Depletion of glu and high NH_3 level drive GDH to reverse: Glu + NAD (P) → αKG + NAD(P) H + NH_4^+ This depletes α KG, an essential TCA cycle intermediate, thus impairing energy metabolism in brain. 3. CPS I deficiency results in inability for nitrogenous wastes (ammonia) to be metabolized *via* urea cycle. 4. Ammonia levels rise, leading to brain damage, coma or death.	• Add α-keto acid analogs of essential amino acids • Liver transplant • Gene therapy *Basis of treatment of hyperammonemia with benzoic acid or phenylacetic acid:* Benzoic acid —Glycine→ hippuric acid Phenylacetic acid —Gln→ Phenylacetyl glutamine

Biochemical Screening Tests

Test	*Compound*	*Interpretation*
Ferric chloride Phenistix	Phenyl pyruvate	Green PKU
Dinitrophenyl hydrazine	α keto acids	Yellow ppt, MSUP, Keto acidosis
Cyanide nitroprusside	Disulfide bonds	Red purple, cystinuria, homocystinuria
Nitrosanaphthol	OH-phenolic acids	Orange red Tyrosinemia liver dysfunction
Merckoquant sulfite	Sulfite	Pink Sulfide oxidase molybdenum decrease
Benedict's (clinitest)	Reducing substance	Green → orange, glycinuria

Acetest	Acetone acetoacetate	Purple, acetoacetate
P nitroaniline	Methylmalonic acid	Emerald green, MMA (methyl malonyluria)
Interpretation of Tests in Metabolic Disorders of Protein Metabolism		
Defect	***Test***	***Interpretation***
Phenylketonuria	Bacterial inhibition assay Alternative methods available: colorimetric, fluorometry and MS/MS (estimate phenylalanine to tyrosine ratio).	Abandoned now as more accurate, faster and efficient methods are available.
Urea cycle disorders	By measuring citrulline, arginine, Ornithine levels; pyroglutamic acid & Glutamine- detected by blood spot assay	•*Citrullinemia*- assessed by measuring citrulline •*Argininosuccidinic aciduria*- assessed by measuring arginine levels •*Ornithine transcarbomylase* (OTC) deficiency- detection of pyroglutamic acid (derived from glutamine) & Glutamine- detected by blood spot assay • *Hyperornithinemia, hyperammonemia, homocitrulinemia and ornithine aminotransferase deficiency*- Ornithine levels assessed
Dicarboxylic amino aciduria	urinary glutamate and aspartate	Defect in renal tubular reabsorption of glutamate and aspartate.

Histidinemia	FIGLU excretion test	FIGLU excretion test
Maple syrup urine disease	Branched chain ketonuria Rothera's test	Branched chain ketonuria results, hence precursors may also appear in urine
Cystathionine beta syntheses deficiency	Methionine levels	diagnosed by elevated methionine levels
Non-ketotic hyperglycinemia	Glycine levels in blood, urine and CSF.	Glycine levels are elevated
Primary Hyperoxaluria	Urinary oxalates	Increased excretion of oxalates is observed up to 600 mg/day
Cystinuria	Urine Cystine, lysine, arginine & ornithine(COLA), microscopy	Increased excretion of Cystine, lysine, arginine & ornithine(COLA). Formation of cystine calculi. Urine shows hexagonal flat crystals
Hartnup disease	Urine chromatography for neutral amino acid	Renal transport defect of tryptophan and neutral amino acid Indoles and other tryptophan degradation products in the urine

Disease	**Defect**	**Mechanism/Features**
Metabolic Disorders of Urea Cycle		
Hyperammonemia I	CPS I	Rare

Hyperammonemia II	OTC	X-linked aversion to high protein diet.
Citrullinemia	ASS	Lethargy, seizures.
AS aciduria	ASase	Friable, tufted hair, fatal in early life.
Hyperargininemia	Arginase	Raised arginine. Arginase deficiency does not result in severe hyperammonemia.
Glycine		
Glycinuria	Defect in transport of glycine and imino acid proline and hydroxy proline	Reabsorption of glycine in renal tubules is defective due to defective transporter. increased excretion of amino acid No clinical abnormalities
Primary hyperoxaluria	Defect in failure to catabolize glycoxalate arising from deamination of glycine	Glyoxalate get oxidized to oxalate causing increased urinary excretion of oxalate Urolithiasis, nephrocalcinosis, recurrent UTI, renal failure, HT
Non Ketotic hyperglycinemia	Deficiency of glycine cleavage complex in 3 of 4 components of complex	Severe mental deficiency Glycine being major inhibitory neurotransmitter may be responsible for neurological complications.
Homocysteine metabolism		
Homocystinuria	Cystathione synthase, Me THF-R Me cobalamin deficiency and synthesis	High levels of homocystine an oxidized product of homocysteine in urine and blood. Thrombosis, mental retardation and eye lesions.
Cystinosis	Lysosomal disorder	Abnormal accumulation of cystine in various organs and tissues of body: kidneys, eyes, muscles, pancreas and brain.

Cytathioninuria	Cystathioninase	Elevated levels of cystathionine in blood and excretion in urine.
Cystinuria	Renal transport defect of cysteine amino acid. No enzyme deficiency.	Associated with Increased excretion of Cystine, lysine, arginine & ornithine (COLA). Formation of cystine calculi. Excretion of cystine increases 20-30 times of normal. Urine shows hexagonal flat crystals.
Hypermethioninemia	SAM synthase	methionine is not converted to S- adenosyl methionine and accumulates in blood.
IMD-Sulfur Amino Acid		

I. cystathionine synthase

II. N5 N10 Me THF

III. low N5 Me THF

IV. low N5 Me THF, defective intestinal absorption

V. Hypermethioninemia Methyl adenosyl transferase

VI. Cystathioninuria

Cystathioninase

Sulfituria

Sulfate oxidase

3 Mer, Pyr, Cys

3 Mer S transferase

Disulfiduria (3 Mer → Pyr)

Meth MAS

Cystinuria *Cystinosis*		
IMD: BCAA		
MSUDMaple syrup urine disease	BCKD (dehydrogenase) AR	Maple syrup: burnt sugar odor ↑ Plasma and urine levels of Leu Ile, Val and their keto acids Evident by end of first week of life Difficulty to take feed Vomiting Lethargy Extensive brain damage: mental retardation. Death by 1st year of life Poor myelination Mechanism: not known Treatment BCAA free – diet
Intermittent BC Ketonuria	Variant of MSUD α Keto decarboxylase defect	Symptoms appear later in life, only intermittently Prognosis by diet therapy favorable.
Isovalerate academia	Isovaleryl CoA dehydrogenase defect	Isovaleryl accumulation – cheesy odor of breath and body fluids.

		Features: Vomiting NM irritability Acidosis Mental retardation Coma precipitated by ingestion of excessive proteins Distinct 'Sweaty Odor'
Methylmalonic aciduria	Hereditary deficiency of methyl malonyl CoA mutase acquired vitamin B_{12} deficiency. Valine metabolism: B_{12} dependent reaction: methyl malonyl CoA mutase defective.	Responsive to pharmacologic doses of B_{12} Methyl malonyl CoA Mutase ↓ B_{12} Succinyl CoA Vomiting, dehydration, seizures, failure to thrive, developmental delay, Propionyl academia. Propionyl CoA accumulates which condenses with OAA to form methyl citrate.
Propionyl CoA carboxylase deficiency	Propionyl CoA carboxylase deficiency in isoleucine metabolism High serum propionyl levels	Propionyl CoA Carboxylase ↓ d-methyl malonyl CoA
AKU (alkaptonuria)	Homogentisate oxidase deficiency	Alkapton, a product of Phe and tyrosine metabolism accumulates. Homogentisate autoxidizes to form dark color pigment that accumulates in various tissues, especially

		cartilages namely joint, nose, ear. Urine darkens on standing.
Isovaleric academia	isovaleryl-CoA dehydrogenase deficiency	Patient usually presents with Vomiting, acidosis, and coma follow ingestion of excess protein. Accumulated isovaleryl-CoA is hydrolyzed to isovalerate and excreted.
IMD: Phe: PKU Phenylketonuria		
I-III	Phenylalanine hydroxylase defect (absent/immature, decreased)	Phenyl alanine is not converted to tyrosine and it accumulates in tissues. By other catabolic routes it is converted to phenyl pyruvate, phenyl lactate and phenyl acetate. All are excreted in urine. Symptoms are mental retardation and convulsions.
IV	Dihydrobiopterine reductase defect	
V	Dihydropiopterine biosynthesis defect	
Tyrosinemia		
I	Fumaryl acetoacetate hydroxylase defect: neonatal tyrosinemia	Tyrosine accumulates in blood due to lack of transaminase. It undergoes other routes of catabolism and converted to p-hydroxy. Phenyl acetate and N-acetyl tyrosine. They are excreted in urine along with tyrosine. Symptoms are skin and eye lesions, mental retardation *etc.*
II	Hepatic Tyrosine transaminase	Oculocutaneous form of the disease: Elevated plasma tyrosine levels, skin and eye lesions.

		Local deposition of cellular tyrosine crystals resulting in inflammatory response. Excessive tearing, redness, pain and photophobia. Mental retardation.
Neonatal tyrosinemia	p-OH phenyl pyruvate hydroxylase deficiency	Tyrosine accumulates in blood and excreted in urine. Para hydroxy phenyl pyruvate conversion to homo gentisate is blocked. Accumulation and excretion of para hydroxy phenyl pyruvate occurs.
AKU (alkaptonuria)	Homogentisate oxidase deficiency	Excretion of homogentisic acid in urine. Urine tuns dark on standing due to polymerization on of oxidative products of homogentisic acid. On exposure to O homogentisic acid is oxidized to quinones. Due to deficiency of enzyme homogentisic acid oxidase homogentisic acid is not converted to maleyl acetoacetate, accumulation of homogentisic acid in blood and excretion in urine occurs. Symptoms are connective tissue pigmentation and arthritis.
Parkinson's disease	Decreased production of dopamine	Common features are tremors difficulty in initiating voluntary movement, mask-like face, shuffling gait. Dopamine is important in substantia nigra to mediate motor activity of brain.

Albinism: Defect in Conversion of Tyrosine to Melanin		
	Tyrosine hydroxylase negative	Lacks all visual pigments, hair bulb fail to convert tyrosine to pigment.
	Tyrosine hydroxylase positive	Have some visible pigment and white-yellow hair some melanocytes present.
	Ocular albinism	AR, X-linked, mechanism: unknown.
Carcinoid syndrome	Argentiffinoma: overproduce serotonin	Hypertension, flushes, respiratory distress, diarrhea, wheezing, increased HIAA in urine. Decreased nicotinic acid causing pellagra, since tryptophan is not available for synthesis of vitamin niacin.
Hyperlysinemias	aminoadipic semialdehyde synthase	Impaired breakdown of lysine result in elevated levels of lysine in the blood and urine. These increased levels of lysine do not appear to have any negative effects on the body.
Histidinemia	AR, liver histidase enzyme is deficient	Presents with mental retardation and delayed speech development. Increased plasma histidine levels & increased excretion of histidine & imidazole pyruvate. FIGLU EXCRETION TEST 5g Histidine given 3 times, 4 hourly to individual. Urine examined 24 hrs after initial dose. Normally < 30 mg of FIGLU excreted/day. These levels increased in FOLIC ACID deficiency.

Maple syrup urine disease	α keto acid decarboxylase is either absent or has decreased activity	Odor of urine maple syrup / burnt sugar. Branched chain ketonuria results, hence precursors may also appear in urine, small amount of OH-amino acids produced reduction of ketoacids.
Xanthurenicaciduria	vit. B or pyridoxine deficiency	kynureninase is less active and conversion of 3-hydroxy kynurenine to alanine and 3-hydroxy anthranilic acid is blocked. Through alternative pathways 3-hydroxy kynurenine is converted to xanthurenic acid and get excreted in urine.
Dicarboxylic amino aciduria	Defect in renal tubular reabsorption of glutamate and aspartate	Increased excretion of urinary glutamate and aspartate. Some asymptomatic, others have mental retardation and hypoglycemia.

QUESTIONS

1. Describe the process of urea synthesis.
2. Write briefly on:
   a. Transamination
   b. Ammonia toxicity
   c. Oxidative decarboxylation
   d. Coenzymes in protein metabolism
   e. Uses of ammonia in body
3. Discuss briefly:
   a. Glyine metabolism

b. One-carbon metabolism

c. Catecholamine synthesis

4. Discuss Fate and functions of:

 a. Methionine

 b. Phenylalanine

 c. Tryptophan

5. Short Notes:

 a. Hart nip

 b. Phenyl ketonuria

 c. Maple syrup urine disease

 d. Homocysteinuria

 e. Glutathione

 f. Creatine synthesis

 g. Albinism

 h. Importance of glutathione.

 i. Alkaptonuria

 j. VMA

 k. Carcinoid syndrome

6. Discuss the process of protein synthesis.

7. Compare protein synthesis in prokaryotes and eukaryotes.

8. Discuss post translational modifications in proteins.

9. Give an account of protein inhibitors.
10. What are zymogens? Give example.
11. Describe the different transporter for amino acids.
12. What is gamma glutamyl cycle/Meister‘s cycle?
13. Write a brief note on protein degradation pathway
14. What is transamination? Give example. Name amino acids which do not undergo transamination.
15. What is oxidative deamination and its importance?
16. How is ammonia detoxified in the brain?
17. Which is more toxic – ammonium ions or ammonia?
18. Name the source of nitrogen in urea cycle.
19. Give an account of disorders associated with urea cycle.
20. Make a list of conditions in which ammonia levels are raised.
21. How urea cycle is linked with TCA cycle?
22. What are the mechanism of ammonia toxicity?
23. Name inherited metabolic disorders of amino acid metabolism.
24. Make a list of transmethylation reactions.
25. Name the specialized products synthesized from glycine.
26. What are the metabolic disorders associated with glycine metabolism?
27. Name Sulphur containing amino acids. Which color reaction gives them positive?
28. Name the disorder(s) associated with branched chain amino acid metabolism.
29. Make a list of disorder diagnosed by odour and color of urine.

30. What are polyamines? Describe their synthesis and importance.

31. Describe synthesis of biogenic amines.

32. What are/is the specialized product(s) derived from phenylalanine and their importance?

BIBLIOGRAPHY

Denise R Ferrier. Lippincott illustrated reviews: biochemistry. 7th Edition. Philadelphia Wolters Kluwer; 2017

Donald Voet, Judith G Voet, Charlotte W Pratt. Fundamentals of Biochemistry. 5th Edition. New York: Wiley; 2016.

Geoffrey L Zubay, Dennis E Vance. Principles of biochemistry. Dubuque, Iowa: William C. Brown; 1995.

Jeremy M Berg, Gregory J Jr Gatto, Lubert Stryer, John L Tymoczko. Biochemistry. 9th Edition. New York: Macmillan International Higher Education: WH Freeman; 2019.

Keith Wilson, John M Walker. Principles and techniques of biochemistry and molecular biology. 7th edition. Cambridge: Cambridge University Press; 2017.

Lehninger A, Nelson D, Cox M. Lehninger principles of biochemistry. New York: Worth Publishers; 2000.

Michael A Lieberman, Rick E Ricer. Biochemistry, molecular biology, and genetics. 7th Edition. Philadelphia, Pa Wolters Kluwer; 2020.

Victor W Rodwell, David A Bender, Kathleen M Botham, Peter J Kennelly, P Anthony Weil. Harper's illustrated biochemistry. 31st edition. New York: Mcgraw-Hill Education; 2018.

CHAPTER 9

Intermediary Metabolism

LEARNING OBJECTIVES:	Keywords:
• Identify the intermediates that interconnect the metabolic pathways. • Illustrate the inter-organ metabolic pathways and tissue specific pathways. • Explain the metabolism of body fuels and energetics in different organs. • Describe the diseases associated with failure in metabolic integration.	Alpha-keto acids, Fasting, Intermediary metabolism, Obesity, Phosphoenol pyruvate, Starvation and fed state, Sugar phosphates.

The central interconnecting metabolic pathway (pathways of synthesis, degradation, and interconversion of important metabolites) common to most cells and organisms are referred to as intermediary metabolism.

INTERMEDIATES INTERCONNECTING THE MAJOR METABOLIC PATHWAYS:			
Sugar Phosphates: Triose-P Tetrose-P Pentose-P, Hexose-P	**α-Ketoacids:** Pyruvate Oxaloacetate [OAA] α-ketoglutarate [α-KG]	**Coenzyme A (CoA) Derivatives:** Acetyl-CoA Succinyl-CoA	**Phosphoenolpyruvate (PEP)**
Three metabolic key crossroads: Glucose-6-phosphate. Pyruvate Acetyl-CoA.			

Integration of organ metabolism in different physiological states	The co-ordination of metabolic activities of different organs serves to support glucose homeostasis and provides a steady supply of glucose to meet the needs of the brain and RBCs. This also helps in the storage of fuel when available in plenty.
Three storage forms of fuel	Glycogen, fat and proteins

Simmi Kharb

Four major tissues with specialized metabolic function	Liver, muscle, adipose tissue and brain. No one tissue can survive metabolically without the other.
Metabolic fuels are used during	***Fed state:*** Fuels used by tissues may be derived from ingested, digested and absorbed food. ***Fasting state:*** Fuels used by tissues are derived from mobilized stores of fuel. ***Starvation:*** Occurs after extended fasting.
Interorgan metabolic pathways	The liver supplies glucose, ketone bodies to other tissues. Adipocytes make FA available to other tissues. The circulatory system transports metabolic fuels, intermediates and waste products among tissues.
Certain metabolic pathways occur in multiple tissues	Cori cycle; Glucose alanine cycle.
Metabolic specialization of organs	Various organs have a different metabolic role and metabolic specialization of organs occur as a result of differential gene expression. All metabolic pathways are under precise regulation to adjust the synthesis and degradation of metabolites to physiological requirements. This is mainly determined by the activity of key enzymes.

Major Organs Involved in the Integration of Metabolism

Organ	***Fuel Storage Pool***	***Mobilized Fuel***	***Conditions***
Liver	Glycogen	Glucose Ketone bodies VLDL-TAG	Fasting, exercise Fasting Fed

Muscle	Glycogen Protein	Lactate Alanine, glutamine	Exercise (intense) Fasting
Adipose	TAG	FFA, glycerol	Fasting (moderate) Exercise

Utilization of Body Fuels in Various Organs

	Liver	***Muscle***	***Adipose***	***Brain***
Function	Maintains blood glucose, makes fat, ketone bodies, stores glycogen	Provides movement: Stores glycogen for its own use; stores proteins.	Manages fat stores	Central control Reliant on glucose for energy
Fed state	Stores glycogen Fat synthesis	Stores glycogen Fat and glucose oxidation Protein synthesis	Stores fat	Uses glucose
Fasting	Glycogen breakdown to make glucose Fat breakdown	Glycogen breakdown Fat Oxidation	Releases stored fat	Uses glucose

ENERGY METABOLISM IN VARIOUS ORGANS

Organ	**Energy Reservoir**	**Preferred Substrate**	**Energy Exported**
Brain	None	Glucose KB (Starvation)	None

Muscle at rest	Glycogen	FA	None
Muscle during Exercise	None	Glucose from glycogen	Lactate
Heart Muscle	Glycogen	FA	None
Adipose tissue	TAG	FA	FA and glycerol
Liver	Glycogen, TAG	Amino acids, glucose FA	FA, glucose, KB

<table>
<tr><td colspan="4">CONDITIONS WHERE METABOLIC INTEGRATION IS IMPAIRED</td></tr>
<tr><td>Obesity</td><td>Accumulation of excess adipose tissue</td><td>The chronic imbalance between energy intake and expenditure</td><td>Genetic and environmental factors are also responsible.</td></tr>
<tr><td colspan="4">Type I Diabetes Mellitus, Insulin Resistance and Type II Diabetes</td></tr>
<tr><td colspan="4">Inherited Diseases of Intermediary Metabolism</td></tr>
<tr><td>Defects in Enzymes of Metabolism of:

Amino acids

Carbohydrates

Fatty acids

Mitochondrial energy metabolism</td><td colspan="2">Amino acidopathies

Organic acidurias

Amino acid transport disorders

Ammonia detoxification disorders

Peptide metabolism (GSH) disorders

Disorders of carbohydrate metabolism and transport</td><td>Disorders of fatty acid oxidation

Disorders of ketogenesis

Mitochondrial disorders

Disorders of cobalamin

Folate metabolism disorders

Disorders of transport or utilization of copper, manganese, iron, zinc.</td></tr>
<tr><td colspan="4">There are a large number of inherited conditions resulting from disorders of intermediary metabolism that result in the built up of precursors, absence of these. Many of these are fatal in childhood, but maybe compatible with adult life and pregnancy. Malignant hypertension and</td></tr>
</table>

plasma pseudocholinesterase deficiency and inherited hematological, endocrine, connective tissue or bone disorders are included in this category. Many diagnostic analytical approaches are available for intermediary metabolites in various diseases.

QUESTIONS

1. Write a short note on metabolism of adipose tissue, RBC and Brain.
2. Explain the basis of clinical manifestations of diabetes mellitus.
3. Write in details about regulation of blood glucose.
4. Describe the utilization of glucose in various organs of the body during fasting and fed state.
5. Differentiate between type I and type II diabetes mellitus.
6. Discuss briefly one carbon metabolism.
7. What are different types of PPAR and their functions?
8. What is LXR?
9. What is nitrogen balance?

BIBLIOGRAPHY

Denise R Ferrier. Lippincott illustrated reviews: biochemistry. 7th Edition. Philadelphia Wolters Kluwer; 2017

Donald Voet, Judith G Voet, Charlotte W Pratt. Fundamentals of Biochemistry. 5th Edition. New York: Wiley; 2016.

Geoffrey L Zubay, Dennis E Vance. Principles of biochemistry. Dubuque, Iowa: William C. Brown; 1995.

Jeremy M Berg, Gregory J Jr Gatto, Lubert Stryer, John L Tymoczko. Biochemistry. 9th Edition. New York: Macmillan International Higher Education: WH Freeman; 2019.

Keith Wilson, John M Walker. Principles and techniques of biochemistry and molecular biology. 7th edition. Cambridge: Cambridge University Press; 2017.

Lehninger A, Nelson D, Cox M. Lehninger principles of biochemistry. New York: Worth Publishers; 2000.

Michael A Lieberman, Rick E Ricer. Biochemistry, molecular biology, and genetics. 7th Edition. Philadelphia, Pa Wolters Kluwer; 2020.

Victor W Rodwell, David A Bender, Kathleen M Botham, Peter J Kennelly, P Anthony Weil. Harper's illustrated biochemistry. 31st edition. New York: Mcgraw-Hill Education; 2018.

CHAPTER 10

Chemistry of Nucleotide

LEARNING OBJECTIVES:	Keywords:
• Define the nucleic acid structure and nomenclature. • Describe the biological significance of nucleotide derivatives. • Summarize the role of nucleic acids in our body.	Nucleoside, Nucleotide, Nucleotide derivatives, Purines, Pyrimidine.

NUCLEIC ACID STRUCTURE	
Nucleotide • Combination of heterocyclic amine, a pentose and phosphoric acid. • Monomeric unit of nucleic acids. • Purine and pyrimidines supply building blocks of nucleic acid. • Also, they are high energy intermediates. • Form part of coenzyme: FAD, NAD, NADP, CoA, SAM. • Have regulatory function: signal transduction, second messenger (cAMP, cGMP). • Free nucleotides/ N-bases: Xanthine, hypoxanthine, uric acid.	**The Numbering of Sugars is Primed** ***Nucleoside*** Pentose sugar added to N_9 or N_1 by βN – glycosidic bond ***Nucleotides*** Linking one or more phosphates with a nucleoside onto 5' and of the molecule through esterification

NAMING CONVENTIONS	Nucleoside
Nucleotides	Purine: end in 'sine' Adenosine, guanosine

Simmi Kharb

Start with nucleotide name and add mono-, di-, or triphosphate to it. Adenosine monophosphate Deoxythymidine diphosphate		Pyrimidine end in 'dine' Thymidine, cytidine, uridine
Nucleotide		**Function**
1.	***Adenosine Nucleotides***	
	ATP	Source of energy
	cAMP	Second messenger
	Active sulfate *(adenosine 3'phosphate* *5'phospho sulfate PAPS)*	Sulfur donor (proteoglycans) Sulfur conjugation of drugs
	Active methionine *(5-adenosyl methionine SAM)*	Methyl donor, source of propylamine, in polyamines
2.	***Guanosine Derivative***	
	GDP	Coupled to substrate-level phosphorylation
	GTP	Allosteric regulation, energy source
	cGMP	Intracellular signal second messenger (NO)
3.	***Hypoxanthine IMP***	Purine salvage pathway
4.	***Uracil***	Glycogen synthesis
	UDP-G	Glycoprotein synthesis

	UDP-Gln	Glucuronide conjugation reaction of bilirubin, drugs
5.	Cytosine	
	CTP	Phosphoglycerate synthesis
	CDP	CDP choline: formation of sphingomyelin with ceramide
II	***Coenzyme***	NAD, FAD, NADP, CoA, SAM
III	***Monomeric precursors***	Monomeric unit of RNA, DNA

Function	**Example**
Energy Metabolism	ATP, muscle contraction, active transport, ion gradient, phosphate donor
Monomeric Unit	NTP, dNTP (for RNA, DNA)
Physiological Mediators	cAMP, cGMP (second messenger) Signal transduction (GTP binding protein) Adenosine (coronary blood flow)
Precursor Function	GTP (mRNA capping)
Activate Intermediates	UDP-G (glycogen) CDP-choline (phospholipid) SAM (methylation) PAPS (Sulfation)
Allosteric Affects	ATP (negative for PFK) AMP (positive for phosphorylase B) dATP (negative effective RNP reductase)

Cofactors	NADP/NADPH, $FAD/FADH_2$, $FMN/FMNH_2$, CoA, SAM

QUESTIONS

1. Discuss the structure of:
   a. Nucleotide
   b. Nucleoside
   c. Purine
   d. Pyrimidine
2. Discuss the role of nucleic acids in energy metabolism.
3. Write a short note on nucleotide analogues.
4. Name the nucleic acids that act as cofactors/ coenzymes, and also provide the corresponding dietary vitamin associated with them.
5. Name the 4 types of nucleotides found in DNA. Also, give the complementary pairs that form Hydrogen bonds in its double helical structure.
6. Explain with examples the biological significance of nucleotides.

BIBLIOGRAPHY

Denise R Ferrier. Lippincott illustrated reviews: biochemistry. 7th Edition. Philadelphia Wolters Kluwer; 2017

Lehninger A, Nelson D, Cox M. Lehninger principles of biochemistry. New York: Worth Publishers; 2000.

Victor W Rodwell, David A Bender, Kathleen M Botham, Peter J Kennelly, P Anthony Weil. Harper's illustrated biochemistry. 31st edition. New York: Mcgraw-Hill Education; 2018.

CHAPTER 11

Nucleotide Metabolism

LEARNING OBJECTIVES:	Keywords:
• Describe the pathways of purine and pyrimidine biosynthesis and degradation. • Define the significance and regulation of nucleic acid metabolism. • Correlate the clinical significance of nucleotide metabolic disorders.	*De novo* and salvage pathways, Gout, Lesch Nyhan Syndrome, One-carbon metabolism, OPRT, PRPP, SCID, THF.

FACTS

Characteristic Features of Purines	**Function**
Purine synthesis occurs in the liver. Nitrogenous bases and nucleotides are then transported to other tissues by red blood cells (RBCs). The brain can also synthesize nucleotides. Purines are not metabolized to central metabolites. The ring structure is retained in birds, reptiles, and primates. Other organisms can further breakdown the ring before excretion.	Provides bases (A and G) for energy metabolism and for DNA and RNA synthesis. **Location** The cytoplasm of most cells. **Regulation** Amidotransferase. Adenosine triphosphate (ATP), guanosine triphosphate (GTP).
Connected to the Following Pathways	***Purine Ring Built-up Using the Following Substances***
Folate metabolism. One-carbon metabolism.	Amino acids glycine, aspartate, glutamine.

Simmi Kharb

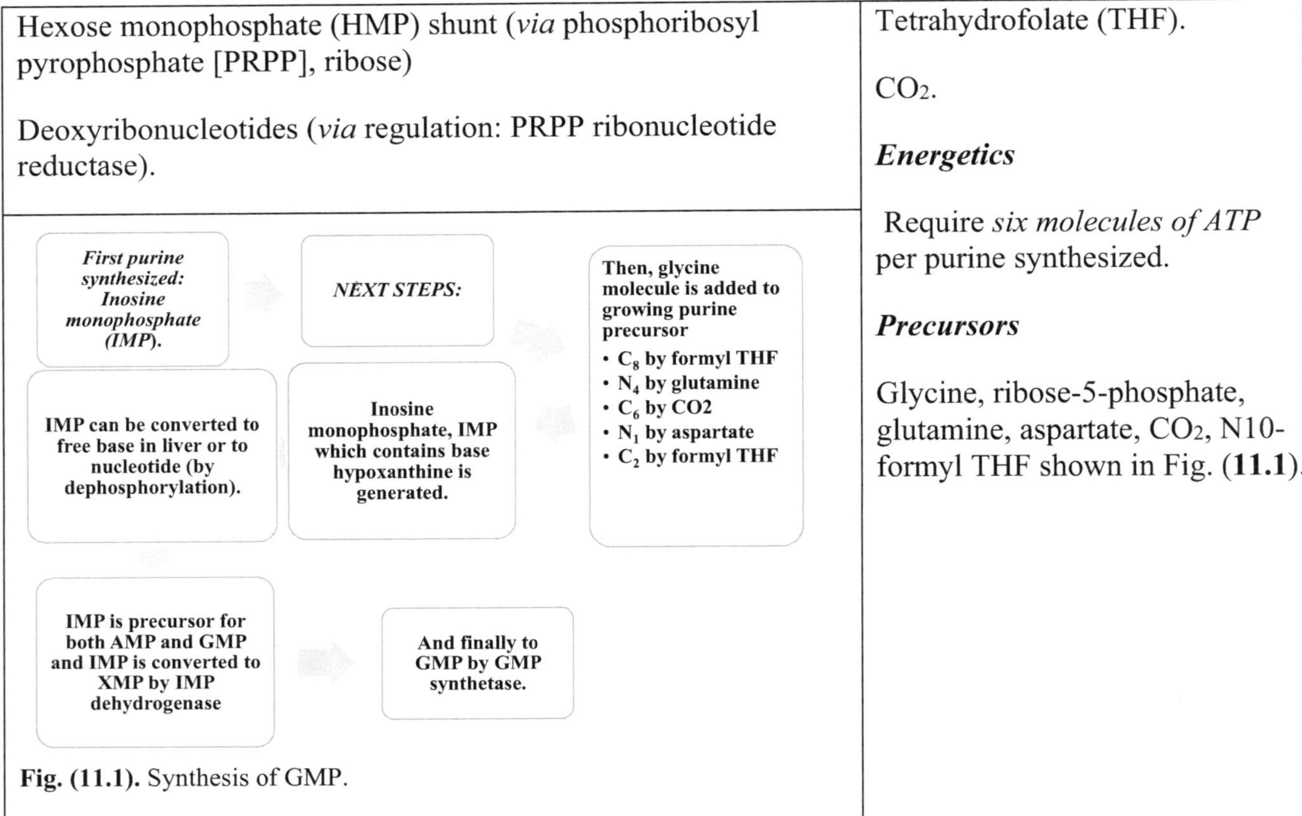

Hexose monophosphate (HMP) shunt (*via* phosphoribosyl pyrophosphate [PRPP], ribose)

Deoxyribonucleotides (*via* regulation: PRPP ribonucleotide reductase).

Fig. (11.1). Synthesis of GMP.

Tetrahydrofolate (THF).

CO_2.

Energetics

Require *six molecules of ATP* per purine synthesized.

Precursors

Glycine, ribose-5-phosphate, glutamine, aspartate, CO_2, N10-formyl THF shown in Fig. (**11.1**).

Purine Synthesis	**Salvage**	***De novo***
Synthesis of purines occurs in two ways: Salvage and *de novo*	Activated ribose (PRPP) + base → nucleotide	Activate ribose (PRPP) + A.A. + ATP + CO_2 → nucleotide
	Base is attached	The base is synthesized (require ATP)
	Preformed bases recovered and reconnected to ribose unit	Nucleotide base assembled first then attached to ribose: pyrimidine In purine: base synthesized piece by piece over the ribose-based structure

1st Step:	PRPP (phosphoribosyl pyrophosphate) synthesis Ribose-5-P +ATP —PRPP synthase→ PRPP + AMP	PRPP provides ribose moiety to glutamine to form PRA (phosphoribosylamine)
	5 phosphoribosyl -1-pyrophosphate ↓ Glutamine → Glutamate ***Phosphoribosyl aminotransferase*** 5 phosphoribosylamine (PRA)	*Aminotransferase:* Committed step Important regulated step Inhibit by IMP, GMP, AMP
Next Steps:	Then, the glycine molecule is added to growing purine precursor C_8 by formyl THF. N_4 by glutamine C_6 by CO_2 N_1 by aspartate C_2 by formyl THF Inosine monophosphate, IMP generation (contains base hypoxanthine)	IMP can be converted to a free base in the liver or to the nucleotide (by dephosphorylation). IMP is a precursor for both AMP and GMP and IMP is converted to XMP by IMP dehydrogenase and finally to GMP by GMP synthetase.
Salvage Pathway	• Requires less energy than denovo • Involves ribosylation of free purine • Occurs in the brain, RBC, polymorphs: →	→ Lack APRT: cannot synthesize PRA and utilize exogenous purines to form nucleotides

Ribosylation of Free Purine by PRPP

1. *Adenine*

AMP + ATP —Adenylate kinase→ 2 ADP

GMP + GTP —Guanylate kinase→ 2 GTP

2. *Hypoxanthine by HGPRT*

IMP → Inosine → Hypoxanthine

PPi

HGPRT

PRPP

GMP → Guanosine → Guanine

3. *Direct Ribosylation of Purine Ribonucleotide*

Adenosine —Adenosine kinase→ AMP

Deoxycytidine —dCK→ dCMP

Deoxy guanosine —dCK→ dGMP

Deoxy adenosine —AK→ dAMP

CATABOLISM

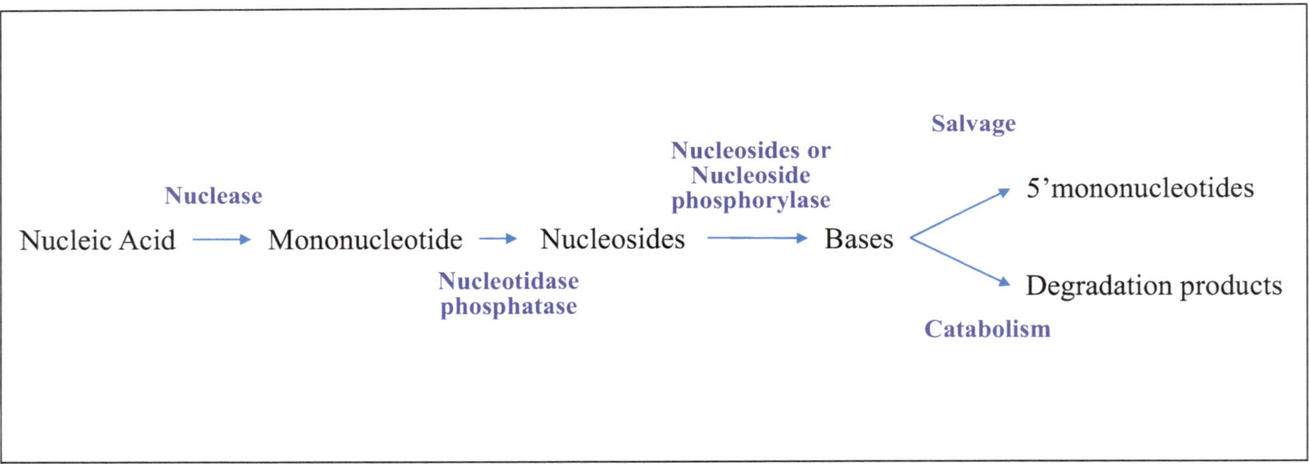

<table>
<tr><td colspan="2">COMMON DISORDER ASSOCIATED WITH NUCLEOTIDE METABOLISM</td></tr>
<tr><td rowspan="2">1. SCID (Deficiency of ADA or PNP)

Leads to SCID (severe combined immunodeficiency):

- dATP and dGTP accumulate

- T-, B- cell and natural killer cell deficiency.</td><td>AMP is degraded to adenosine, by RNTDase, then to inosine finally by adenosine deaminase enzyme.

Then, xanthine is converted to uric acid by xanthine oxidase.</td></tr>
<tr><td>Inosine is degraded by PNP to hypoxanthine and R-1-P. Hypoxanthine is oxidized to xanthine by xanthine oxidase.</td></tr>
<tr><td colspan="2">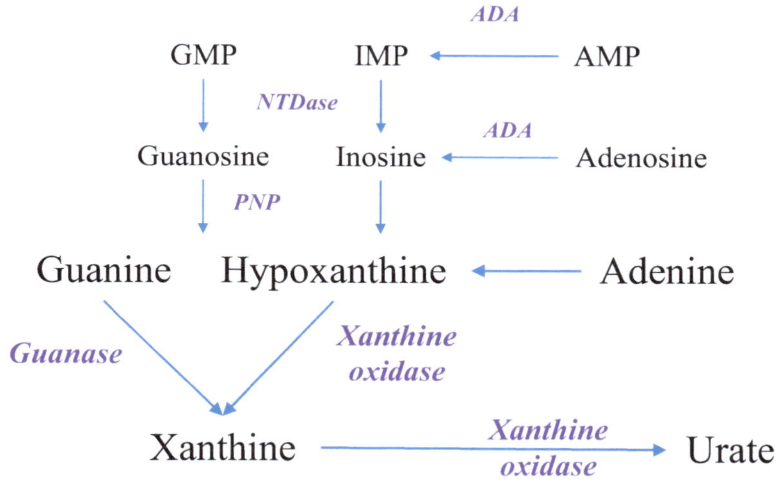</td></tr>
<tr><td>2. GOUT</td><td>Peripheral arthritis resulting from deposition of sodium urate crystals in one or more joints.</td></tr>
</table>

Mechanism R-5-P —||→ (*PRPP-S*) [PRPP] —||→ PRA ——→ IMP —||→ AMP; IMP —||→ GMP			Not inhibited by IMP, AMP, GMP
Defect	i. PRPPS enzyme defect	Superactive (increased Vmax) Enzyme resistant to feed back inhibition Low Km for R-5-P.	
	ii. HGPRT deficiency	PRPP levels increased.	
Features	Hyperuricemia	Attacks of actuate, monoarticular, inflammatory arthritis. Tophaceous deposition of urate crystals in and around joints Intestinal deposition of urate crystals in renal parenchyma, urolithiasis.	
Pathophysiology	Primary gout	Overproduction: 10% Under excretors: 90%	
	Secondary gout	Excess nucleoprotein turnover (lymphoma, leukemia) Increased cell proliferation death (psoriasis)	
Treatment	NSAID (non-steroidal anti-inflammatory drugs)	Prednisone ACTH	

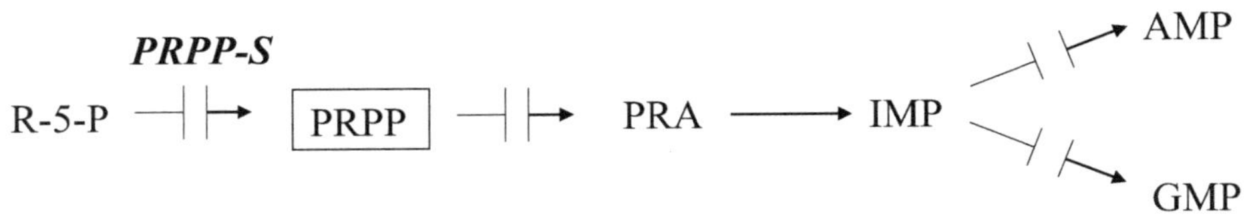

	Colchicine Intrarticular corticosteroid	Opiates
3. **Adenosine Deaminase (ADA) Deficiency**	***ADA*** Adenosine⟶Inosine	Autosomal recessive disorder, infant prove to bacterial, candida, viral, protozoal infections Both T and B-cells significantly reduced.
4. **Lesch Nyhan Syndrome**	***Defect in Production or Activity of HGPRT***	Increased level of hypoxanthine and guanine Increased degradation to uric acid PRPP accumulates Causes gout like symptoms Neurological symptoms: Spasticity, aggressiveness, Self-mutilation.
5. Orotic Aciduria	Caused by defect in enzyme UMP synthase complex (OPRT, ODC). Pyrimidines cannot be synthesized→	→ Increased excretion of orotic acid in urine. Retarded growth, severe megaloblastic anaemia. Oral administration of uridine bypasses the metabolic block and provides a source of pyrimidine.

PYRIMIDINE SYNTHESIS	Pyrimidines are not synthesized as nucleotide derivatives; instead, pyrimidine ring is constructed before a ribose-5-P moiety is attached.

Characteristics of Pyrimidine Synthesis	
Requirement Amino acid: glutamine, aspartate; CO_2.	***Location***

<table>
<tr><td>Function

Makes pyrimidine nucleotides (U, T, C) for DNA and RNA synthesis.</td><td rowspan="2">Cytoplasm of most cells: most of pathway takes place in cytoplasm, only one reaction

Takes place in mitochondria (orotate dehydrogenase [ODH]).

Connected to amino acid metabolism (for glutamate and aspartate).</td></tr>
<tr><td>Regulation

UTP, carbamoyl phosphate synthetase (CPS II).</td></tr>
</table>

***De novo* Synthesis of Pyrimidine**	Glutamine + CO_2 + Aspartate + PRPP + ATP → UTP + CTP
Requires six enzymatic steps. These enzymes exist as two enzyme complexes:	CAD (CPS II, ATCase, DHOase) Uridine monophosphate (UMP) synthase (OPRT, orotidine monophosphate decarboxylase [OMP-DC]).

Glutamine combines with CO_2 and ATP in the presence of ***CPS II*** (in cytosol) to form carbamoyl phosphate.

$$\text{Carbamoyl phosphate + Aspartate} \xrightarrow[\textit{Aspartate Transcarbamoylase}]{\textit{ATCase}} \text{N-carbamoyl aspartate}$$

Transcarbamoylase Step Entire aspartate combines with carbamoyl phosphate catalysed by transcarbamoylase enzyme to produce dihydroorotate.	Committed step Stimulated by ATP, PRPP, and dihydroorotate Inhibited by UMP.
Ring Closure and Dehydration	Orotate is formed through the action of *dihydroorotase.*

Orotate phosphoribosyl transferase (OPRT) catalyzes transfer of ribose-5-phosphate from PRPP to form 5′-orotate monophosphate (5′-OMP).

OMR-DC (OMP decarboxylase) Step	OMP decarboxylated to form UMP by OMP decarboxylase

UMP is synthesized first CTP is synthesized from UMP Two kinases reactions convert UMP to UTP:	UMP + ATP → UDP + ADP UDP + ADP → UTP + ADP
UMP is Aminated to CTP	H_2O Gln, ATP ADP+Pi Glu UTP → CTP ***CTP synthetase***
CDP is reduced to dCDP by ***ribonucleotide reductase*** which is deaminated to dUMP.	dUMP is converted to dTMP by ***thymidylate synthase***
Interrelationship Between Urea and Pyrimidine Synthesizing Pathway	
Initial Reactant in Both the Pathways Carbamoyl phosphate	Two carbamoyl phosphate pools are compartmentalized: CP produced by liver mitochondria may become available for utilization in de novo pyrimidine biosynthesis: ***Applied*** In hyperammonemia due to defect in urea cycle enzyme (OTC): There is accumulation of CP and patient may excrete orotic acid, uridine, and other pyrimidine-containing compounds.
Enzymes Mitochondrial CPS (I) in urea synthesis Cytosolic CPS (II) in pyrimidine biosynthesis	
Orotic Aciduria can Occur in Both Hereditary orotic aciduria (block of phosphoribosyl transferase and orotidylate decarboxylase steps) and Urea cycle defects with hyperammonemia.	
CATABOLISM OF PYRIMIDINES	Pyrimidines are degraded in liver to CO_2 and β-alanine and some nitrogen are released as ammonium ion

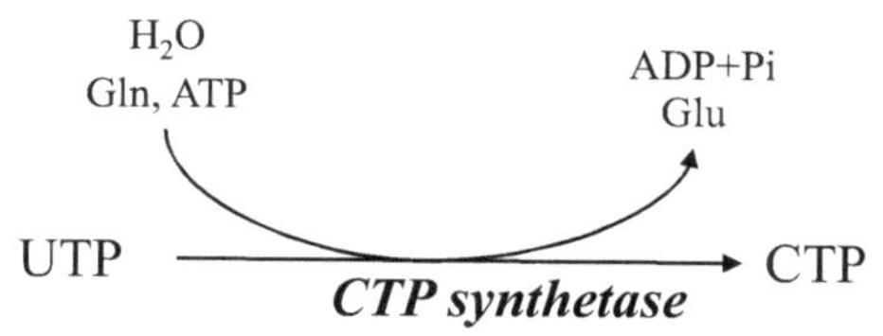

It Requires Enzymatic Steps: - Dihydrouracildehydrogenase - Dihydropyrimidine - Ureidopropionase	**Pyrimidine Catabolism Results in:** Degradation of pyrimidine ring to water-soluble products: - β-alanine, β-aminoisobutyrate
Through Transamination and Activation Reactions β-alanine and β-aminoisobutyrate → malonyl-CoA and methylmalonyl-CoA.	
Malonyl-CoA	Precursor of fatty acid (FA) synthesis Converted to succinyl-CoA (TCA cycle intermediate: contributes to energy metabolism)
β-Alanine	Incorporated into carnosine and anserine and excess is excreted in urine Urinary excretion serves as a marker in lymphoproliferative disorders
Salvage Pathway of Pyrimidine	***Two Types***
Minor pathway Pyrimidine nucleosides from exogenous sources and from endogenous breakdown of nucleic acids are utilized for synthesis of pyrimidine nucleotides:	***<u>First Pathway Involves</u>*** Attachment of base to PRPP with formation of pyrophosphate This pathway occurs with purine and uracil but not for cytosine or thymine
Uridine kinase Uridine ----------------→UMP ***Cytidine kinase*** Cytidine -----------------→CMP	***<u>2nd pathway involves:</u>*** Attachment of base to ribose-1-phosphate Occurs for most of purines and pyrimidines

Thymidine Thymidine ---------------→ TMP ***Kinase***	Requires presence of kinases (which convert nucleosides to monophosphate)
dC-kinase D-cytidine -----------------→ dCMP	***OPRT*** OA ------------→OMP
NUCLEOTIDE ANALOGS	
Azidothymidine (AZT):	Used in treatment of AIDS.
Dideoxy cytidine (ddC):	Antiviral, stops DNA or RNA synthesis directly.

QUESTIONS

1. Differentiate between CPS I and CPS II.
2. Write your comments on : Genetic code: features and properties
3. What are the sources of one-carbon?
4. What is purine salvage pathway and its importance?
5. Explain the biochemical basis of gout.
6. Write a short note on:
   a. Lesch Nyhan syndrome
   b. nucleotide analogues
   c. SCID
   d. One-carbon metabolism
   e. Gout
   f. Orotic acid urea

7. Discuss synthesis of purine briefly by drawing diagram and first two steps.

8. Describe metabolic uses of nucleic acids briefly. Which nucleic acids are used as therapeutic agents

BIBLIOGRAPHY

Denise R Ferrier. Lippincott illustrated reviews: biochemistry. 7th Edition. Philadelphia Wolters Kluwer; 2017

Donald Voet, Judith G Voet, Charlotte W Pratt. Fundamentals of Biochemistry. 5th Edition. New York: Wiley; 2016.

Geoffrey L Zubay, Dennis E Vance. Principles of biochemistry. Dubuque, Iowa: William C. Brown; 1995.

Jeremy M Berg, Gregory J Jr Gatto, Lubert Stryer, John L Tymoczko. Biochemistry. 9th Edition. New York: Macmillan International Higher Education: WH Freeman; 2019.

Lehninger A, Nelson D, Cox M. Lehninger principles of biochemistry. New York: Worth Publishers; 2000.

Michael A Lieberman, Rick E Ricer. Biochemistry, molecular biology, and genetics. 7th Edition. Philadelphia, Pa Wolters Kluwer; 2020.

Victor W Rodwell, David A Bender, Kathleen M Botham, Peter J Kennelly, P Anthony Weil. Harper's illustrated biochemistry. 31st edition. New York: Mcgraw-Hill Education; 2018.

CHAPTER 12

Molecular Biology

LEARNING OBJECTIVES:	Keywords:
• Describe the characteristics, structure, types and functions of DNA and RNA. • Explain cell cycle events and replication of DNA in prokaryotes and eukaryotes. • Appraise process and role of transcription and translation in cells. • Summarize the significance of genetic code and mutations in organisms. • Illustrate the regulation of gene expression. • Identify the basis of cancer and apoptosis.	Apoptosis, Cell cycle, Chargaff's Rule, Cancer, Caspases, DNA, Gene, Mutation, Operons, Oncogene, Polymerases, RNA, Replication, Reverse transcriptase, Ribosomes, Translation, Transcription.

12.1. DNA AND RNA STRUCTURE

Structure of DNA	DNA is composed of a nitrogen base (A, G, C, T) deoxyribose and phosphates.
DNA Double Helix • Double stranded helix with major and minor grooves • Composed of two polynucleotide chains joined by hydrogen bonds between bases • Chains are antiparallel: Chain runs in 5' to 3' direction and other in 3' to 5' • In the interior of molecule base pairs of strands are stacked (like a spiral staircase)	Four nitrogenous bases: • Purines: A, G • Pyrimidines: C, T • Adenine base pairs with thymine (two bonds) • Guanine base pairs with cytosine (three bonds)
	Base pairing on two strands are complementary:

Simmi Kharb

<table>
<tr><td colspan="2">and phosphate groups are on the outer side of the double helix.</td><td colspan="2" rowspan="2">• PURINE always pairs with PYRIMIDINE
• Adenine on one strand base pairs with thymine on other strands
• Guanine on one strand base pairs to cytosine</td></tr>
<tr><td colspan="2">Chargaff's Rule
- concentration of [A] = [T]; [C] = [G]</td></tr>
<tr><td colspan="4">Structure of RNA</td></tr>
<tr><td colspan="2">RNA is usually a single-stranded molecule.
Three types of RNA namely, mRNA, rRNA and tRNA:</td><td colspan="2">RNA differs from DNA since RNA contains ribose sugar instead of deoxyribose and uracil (U) rather than thymine.</td></tr>
<tr><td>mRNA
(messenger RNA)</td><td colspan="3">Contains a cap (of methylated GTP) and poly A tails
mRNA is a source of coding information for protein synthesis</td></tr>
<tr><td>rRNA
(ribosomal RNA)</td><td colspan="3">Contains many loops and base-pairing
They are associated with proteins to form ribosomes</td></tr>
<tr><td>tRNA
(transfer RNA)</td><td colspan="3">Act as adaptors that can bind an amino acid at one end and interact with mRNA at other They have a clover leaf structure and contain modified nucleotides
The first loop of clover leaf at 5’ end comprising of dihydrouridine; middle loop contains anticodon which base pairs with a codon in mRNA and third loop are TψC containing both ribothymidine and pseudouridine
At 3’ end, the CCA sequence carries the amino acid</td></tr>
<tr><td colspan="2">REPLICATION</td><td colspan="2">Cell Cycle Phases</td></tr>
<tr><td colspan="2">Cell Cycle in Eukaryotes
It is a cell’s lifetime of growth and division (Fig. 12.1).</td><td>S Phase</td><td>• DNA synthesized
• Eukaryotes replicate only once per cell division cycle</td></tr>
</table>

Regulated by cyclins and kinases. When cell cycle checkpoints are broken or the cell cycle goes wrong, cells may undergo apoptosis or become cancerous.		• Lasts for 8 hrs in humans
	G_2 and M phase	• Gap phase G_2 occurs before mitosis and cell division (M phase) • The variable time between two phases
Applied Paclitaxel (Taxol) prevents cytoskeleton remodeling during cell division: Used in cancers.	*G_1 phase*	• Normal cellular functions take place during this phase
	G_0 phase	• Resulting phase • Variable time • Tumor cells do not enter the G_0 phase
Semi Conservative Replication	Process of making an identical copy of a portion of DNA, using existing DNA as a template for synthesis of new complementary DNA strand	

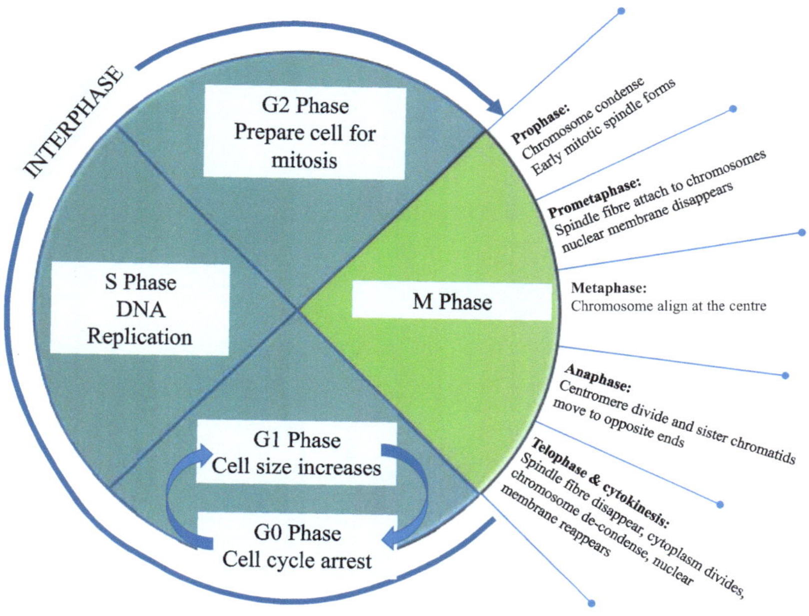

Fig. (12.1). Phases of the Cell Cycle.

Facts

S phase of Interphase

Occurs in a semi-conservative manner

Each strand of ds parental DNA molecule:

- Separates from its complement during replication
- Serves as a template on which a new complementary strand is synthesized

Basic Requirements for DNA Synthesis'				
Substrate NTPs	***E***	***Template***	***Primer***	***Enzyme***
dATP, dGTP, dCTP, dTTP	Cleavage of high energy phosphate bond between α and β phosphate	Each strand of parental DNA	To prepare template strand for addition of nucleotide	DNA-dependent DNA polymerase

Proteins Involved in DNA Synthesis (Fig. 12.2)						
DNA polymerases	DNA Helicases	DNA Primases	Single-stranded DNA binding proteins	DNA ligase	Topoisomerases	Telomerase

Three Steps	***Initiation***	***Elongation***
Initiation Elongation Termination.	Begins at origins of replication Bidirectional with most genomes *Priming with RNA:* Requires DNA primase enzyme	*Three steps:* Unwinding of DNA Relieving of torsional stress Polymerization of a polynucleotide chain One strand of DNA is made continuously while the other strand is synthesized in fragments.

Elongation:	*Unzipping:*	*Topoisomerases*:	*Single-stranded DNA binding proteins*:	*DNA polymerases:*
	Helicase unzips the genes of interest	Relax the super coiling of DNA		Add and read proof new bases

<table>
<tr><td rowspan="3"></td><td>Recoils the two, new identical DNA molecules.</td><td>Primases:

Synthesize a prime.</td><td>Prevent premature annealing of ssDNA to double-stranded DNA.</td><td rowspan="3">Okazaki Fragments:

Short, newly synthesized DNA fragments formed on lagging template strand during DNA replication.

Replication fork:

Y-shaped region formed by unwinding of parent strands.</td></tr>
<tr><td>Ligases:

Link the added bases by catalyzing the formation of a phosphodiester bond.</td><td>Synthesis:

DNA molecules synthesized by assembling nucleotides, building blocks of DNA.</td><td rowspan="2">Leading Strand: Strand of DNA is replicated continuously.

Lagging Strand: Strand of DNA whose direction of synthesis is opposite to direction of growing replication fork.</td></tr>
<tr><td colspan="2">Proofreading: DNA polymerase</td></tr>
<tr><td colspan="2">Origins of Replication</td><td>Where replication of DNA molecule begins.</td><td>Bacterial chromosome: circular, single origin.</td><td>Eukaryotic chromosome: linear, thousands of origins.</td></tr>
</table>

<table>
<tr><td colspan="3">DNA Polymerases</td><td>Function</td><td colspan="2" rowspan="2">Replication of DNA

Repair of damaged DNA</td></tr>
<tr><td colspan="4">Types of DNA Polymerases</td></tr>
<tr><td rowspan="2">E. coli: 3 types: I, II, III</td><td><u>Types</u></td><td colspan="3"><u>Features</u></td><td><u>Functions</u></td></tr>
<tr><td>Pol I</td><td colspan="3">Used to fill gap between DNA fragments of lagging strand

Major enzyme for gap-filling during DNA repair</td><td>Proof reading

Excision-repair

Nick translation</td></tr>
</table>

	Pol II	Encoded by Pol B gene, involved in SOS response to DNA damage	DNA repair Proof reading function (3'→ 5') Lacks excision repair activity.
	Pol III	Carries out DNA replication	Catalyzes leading and lagging strand synthesis Proof- reading 3'→5' action No excision repair action

Eukaryotic DNA Pol	
DNA pol ∈	Synthesize leading strand during replication
DNA pol δ	Synthesize lagging strand
DNA pol γ	Replicates mitochondrial DNA
DNA pol σ	Primase
DNA pol β and ∈	Participate primarily in DNA repair
DNA pol α	Replication of nuclear DNA

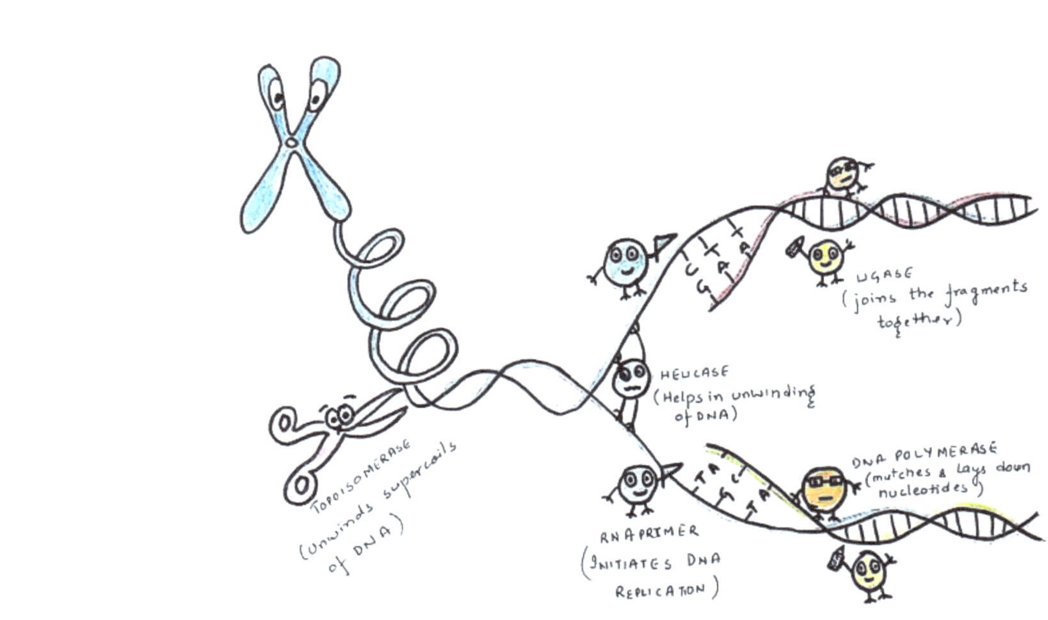

Fig. (12.2). DNA Replication.

Comparison

	Step in DNA replication	***Prokaryotes***	***Eukaryote***
1	Recognition of origin of replication	dnaA protein	-
2	Unwinding of DNA double helix	Helicase, requires ATP	Helicase requires ATP
3	Synthesis of RNA primers	Primase	Primase
4	Synthesis of DNA:		
	Leading strand	DNAPol III	Pol$\in$
	Lagging strand	Pol III	Polδ
	Okazaki fragment		

5	Removal of RNA primer	DNA pol I (5'→3' exonuclease)	Unknown
6	Replacement of RNA with DNA	DNA Pol I	Unknown
6	Joining of Okazaki fragments	DNA ligase (require NAD)	DNA ligase (require ATP)
7	Removal of positive supercoils ahead of advancing replication fork	DNA topo I (DNA gyrase)	DNA topo II
8	Synthesis of telomeres	Not required	Telomerase

Remember: DNA polymerase can only copy a DNA template in 3' → 5' direction and produce newly synthesized strand in 5' to 3' direction.

Telomerase	• Bacterial DNA is circular: do not have end-replication problem • In eukaryotes: chromosome ends are called telomeres: ***Two Main Functions*** • To protect chromosomes from fusing with each other • To solve the end-replication problem	***Telomeres*** • Short tandem repeats of sequence of TTAGGG • 3' end of these repeats is single stranded • Present at the end of a chromosome, shortens with every cell division • Telomere shortening is recognized as and is a part of normal aging process

<table>
<tr><td>DNA Telomerase:</td><td>Specialized DNA polymerase in eukaryotes used to maintain telomeres:

Replicates telomeric ends

Contains a short RNA template complementary to DNA telomere sequence and telomerase reverse transcriptase activity (htrt)

Able to replace telomere sequences that would otherwise be lost during replication</td><td>Clinical Relevance

Normally telomerase activity is present only in embryonic cells, germ (reproductive) cells, and stem cells, but <u>not</u> in somatic cells

Cancer cells often have relatively high levels of telomerase, preventing telomeres from becoming shortened and contributing to immortality of malignant cells

Contributes to aging of cells</td></tr>
</table>

Reverse Transcriptase: RNA- Dependent DNA Polymerase	Eukaryotic cells also contain reverse transcriptase activity:
• Requires RNA template to synthesize new DNA • Retroviruses (*e.g.* HIV), use this enzyme to replicate their genomes • Inhibited by AZT, ddc in retroviruses	○ Associated with telomerase ○ Encoded by retrotransposons: residual viral genomes permanently maintained in human: ○ They play a role in amplifying certain repetitive sequences in DNA

DNA Damage (Fig. 12.3)	**DNA Repair**
***Causes*:** • Chemicals or radiation exposure • Incorporation of incorrect base pairs during replication	• Multiple repair systems in prokaryotes and eukaryotes to repair damaged DNA before mutants become fixed by replication. • If cells are allowed to replicate using damaged template, there is risk of:

<table>
<tr>
<td>

<u>*Diseases Associated with DNA Repair*</u>

Inherited mutations that result in defective DNA repair mechanism are associated with a predisposition to development of cancer.

Examples:

Xeroderma pigmentosum

Hereditary non polyposis colorectal cancer (HNPCC).

</td>
<td>

- Introducing stable mutation in new DNA
- Developing cancer if repair mechanism is defective.

DNA Repair Mechanisms

Base excision

Nucleotide excision

Mismatch repair

</td>
</tr>
</table>

	Damage	**Cause**	**Recognition/Excision Enzyme**	**Repair Enzyme**
1.	Thymine dimers (G_1)	UV radiation	Excision endonuclease (deficient in Xeroderma pigmentosum)	DNA Pol DNA ligase
2.	Cytosine deamination (G_1)	Spontaneous/ Chemicals	Uracil glycosylase AP endonuclease	DNA pol DNA ligase
3.	Apurination or Apyrimidination (G_1)	Spontaneous/ heat	AP endonuclease	DNA pol DNA ligase
4.	Mismatched base (G_2)	DNA replication errors	Mutation on one of two genes: hMSH 2 or hMSH 1 Initiates defective repair of DNA mismatches.	DNA pol DNA ligase

			Causes: hereditary nonpolyposis colorectal cancer (HNPCC)	

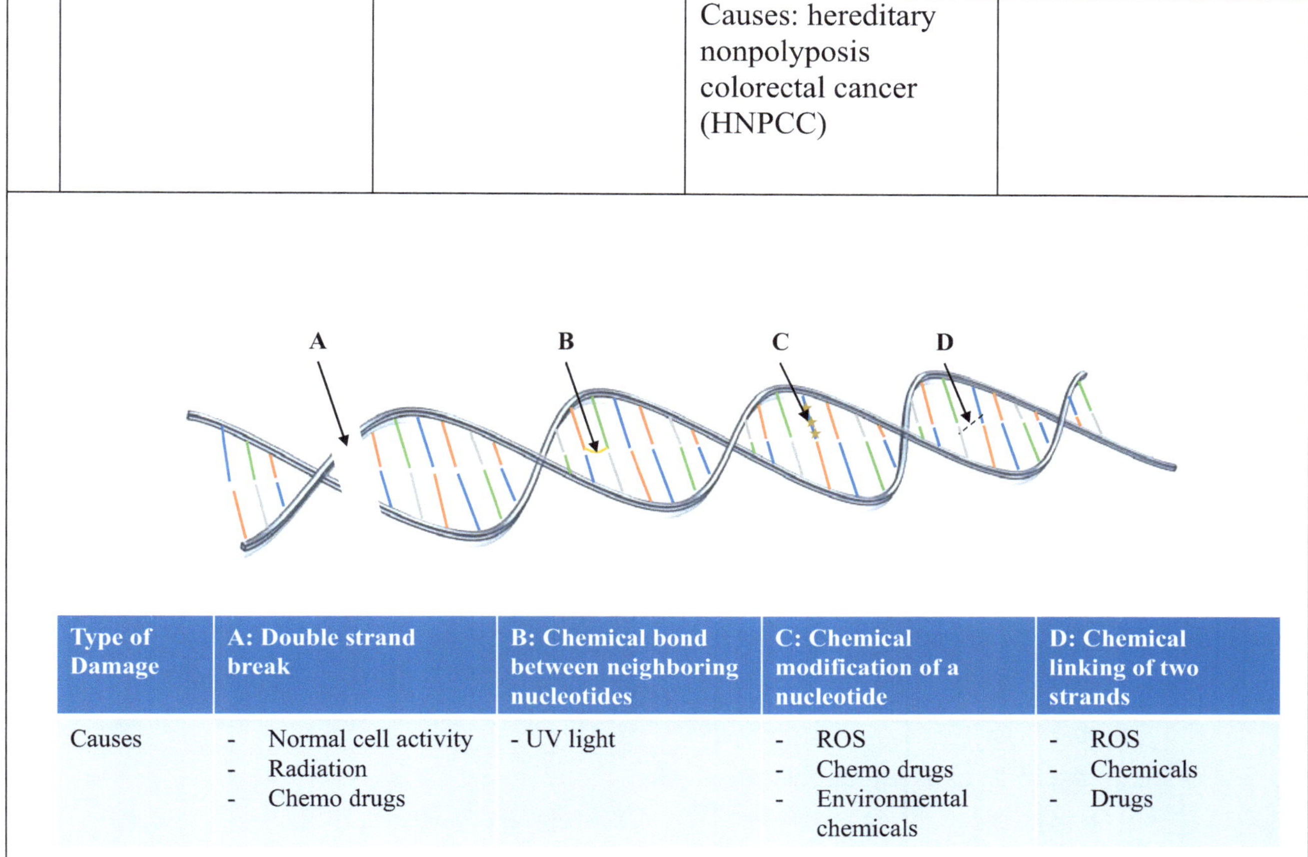

Type of Damage	A: Double strand break	B: Chemical bond between neighboring nucleotides	C: Chemical modification of a nucleotide	D: Chemical linking of two strands
Causes	- Normal cell activity - Radiation - Chemo drugs	- UV light	- ROS - Chemo drugs - Environmental chemicals	- ROS - Chemicals - Drugs

Fig. (12.3). Causes of DNA Damage.

12.3. TRANSCRIPTION

Transcription - Synthesis of RNA	**They Include**
Process of RNA synthesis Controlled by the interaction of promoters and enhancers Several different types of RNA are produced	Messenger RNA (mRNA): specifies the sequence of amino acids in the protein product Transfer RNA (tRNA) and ribosomal RNA (rRNA),: play a role in the translation process.

RNA Differs from DNA	
• RNA exists as single strand, DNA exists as double stranded helix • RNA can be hydrolyzed by alkali to 2', 3' cyclic diesters with mononucleotides, while DNA cannot be hydrolyzed due to absence of 2' hydroxyl group.	• In RNA, sugar moiety is ribose rather than 2' deoxyribose of DNA • Pyrimidine component differs: RNA contains A, G, C, U DNA contains A, G, C, T
Transcription: Synthesis of RNA directed by a DNA template.	***RNA Polymerase in Prokaryotes***
RNA Polymerase: Fact file • RNA polymerase initiate synthesis of new chains. • No primer is required in transcription • DNA copied in 3' to 5' direction RNA chain grows in 5' to 3' direction • Ribonucleotide triphosphate: ATP, GTP, UTP, CTP serve as precursors for RNA synthesis	RNA pol of E coli has four subunits: $\alpha_2\ \beta\beta'$ Forms core enzyme and fifth subunit sigma (σ) for initiation of RNA synthesis. ***Applied aspect*** *Rifampicin:* • Inhibits β subunit of bacterial DNA-dependent RNA pol • Used in treatment of tuberculosis

Eukaryotic RNA pol	Three: I, II, III		
RNA pol	***Location***	***Product***	***β Amantin Sensitivity***
RNA pol I	Nucleolus	285, 185, 5.85 rRNA	Insensitive
RNA pol II	Nucleoplasm	hnRNA/mRNA, SnRNA	Very sensitive
RNA pol III	Nucleoplasm	tRNA, ssRNA, SnRNA	Less sensitive

<table>
<tr><td colspan="5">Comparison RNA pol (RNAP)</td></tr>
<tr><td colspan="3">Prokaryotic</td><td colspan="2">Eukaryotic</td></tr>
<tr><td colspan="3">Single

Synthesize rRNA, except ssrRNA</td><td colspan="2">Three

RNAP I : rRNA except ssrRNA

RNAP II: hnRNA/mRNA, some SnRNA

RNAP III: tRNA, ssrRNA</td></tr>
<tr><td colspan="3">Require sigma factor for initiation</td><td colspan="2">No sigma factor required, require transcription factors (TF II D) which bind before RNAP</td></tr>
<tr><td colspan="3">Sometimes require rho factor to terminate</td><td colspan="2">No rho factor required</td></tr>
<tr><td colspan="3">Inhibited by rifampin, actinomycin D</td><td colspan="2">RNAP II inhibited by amantin (mushrooms) actinomycin D</td></tr>
<tr><td colspan="3">Ribosomes

Site of protein synthesis: mRNA used as template

tRNA causes pairing with appropriate amino acids.</td><td>3 Types</td><td>mRNA

rRNA

tRNA</td></tr>
<tr><td colspan="3">Ribosome in Prokaryotes (Fig. 12.4)

e.g. E.coli ribosome: 25 nm diameter and 2520 kDa in mass

Has 2 unequal subunits: 30S, 50S</td><td colspan="2" rowspan="2">Ribosome in Eukaryotes (Fig. 12.4)

Mitochondrial ribosome similar to prokaryotic

Cytoplasmic ribosomes:

Larger and more complex

40S subunits: 32 proteins 18 S rRNA</td></tr>
<tr><td>30S:</td><td>930 kDa with 20 proteins</td><td>- 16 S rRNA</td></tr>
</table>

50S:	1590 kDa with 33 proteins	- Two rRNA: 23S rRNA, 5S rRNA	60S subunits: 47 proteins 28 S, 5 S and 5.8 S rRNA

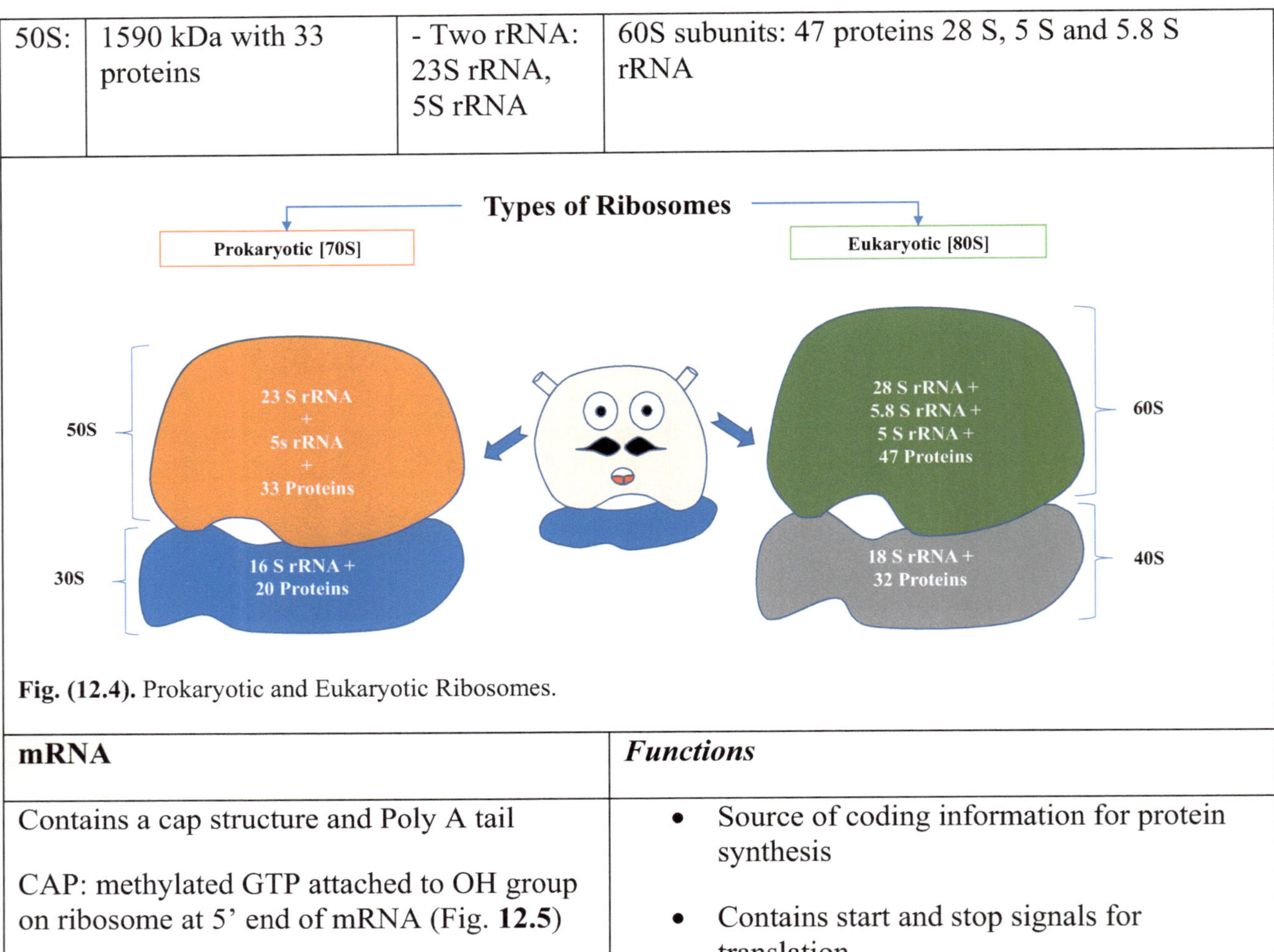

Fig. (12.4). Prokaryotic and Eukaryotic Ribosomes.

mRNA	***Functions***
Contains a cap structure and Poly A tail CAP: methylated GTP attached to OH group on ribosome at 5' end of mRNA (Fig. **12.5**) N^7 of guanine methylated 2'-OH group of 1st and 2nd Ribose moiety may also be methylated.	• Source of coding information for protein synthesis • Contains start and stop signals for translation • Eukaryotic mRNA capped

tRNA	***Structure: tRNA***
Act as adaptors that can bind an amino acid at one end and interact with mRNA at the other.	• Clover leaf structure • Contains modified nucleotides

Each tRNA has: • An amino acid attached to its 3' end • A unique anticodon: can attach to only one amino acid • Serves to transfer amino acids to ribosome	• In eukaryotes, many nucleotides are modified: *pseudouridine (ψ), dihydrouridine (D), ribothymidine (T).* • *1st loop* from 5' end : D loop contains dihydrouridine • Middle loop contains anticodon: base pairs with codon in mRNA • *Third loop: TψC loop*: contains ribothymidine and pseudouridine • CCA sequence at C' end : carries amino acid

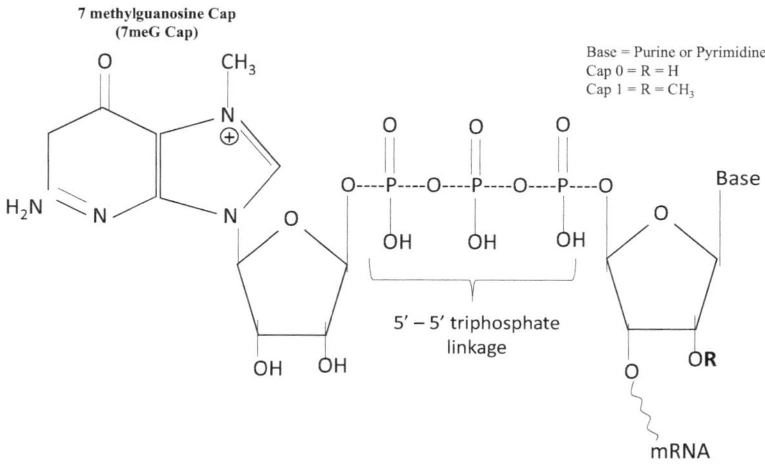

Fig. (12.5). 7-meG cap on mRNA.

Prokaryotic Gene: Function (Fig. 12.6)

***Promoter*:**	***Pribnow Box:***	***Ribosome Binding Site:***	***Coding Region*:**
RNAP binds	TATA AT at-10 -35 site: sigma recognition	Shine Dalgarno	produce amino acid sequence in protein Begins with AUG for methionine.

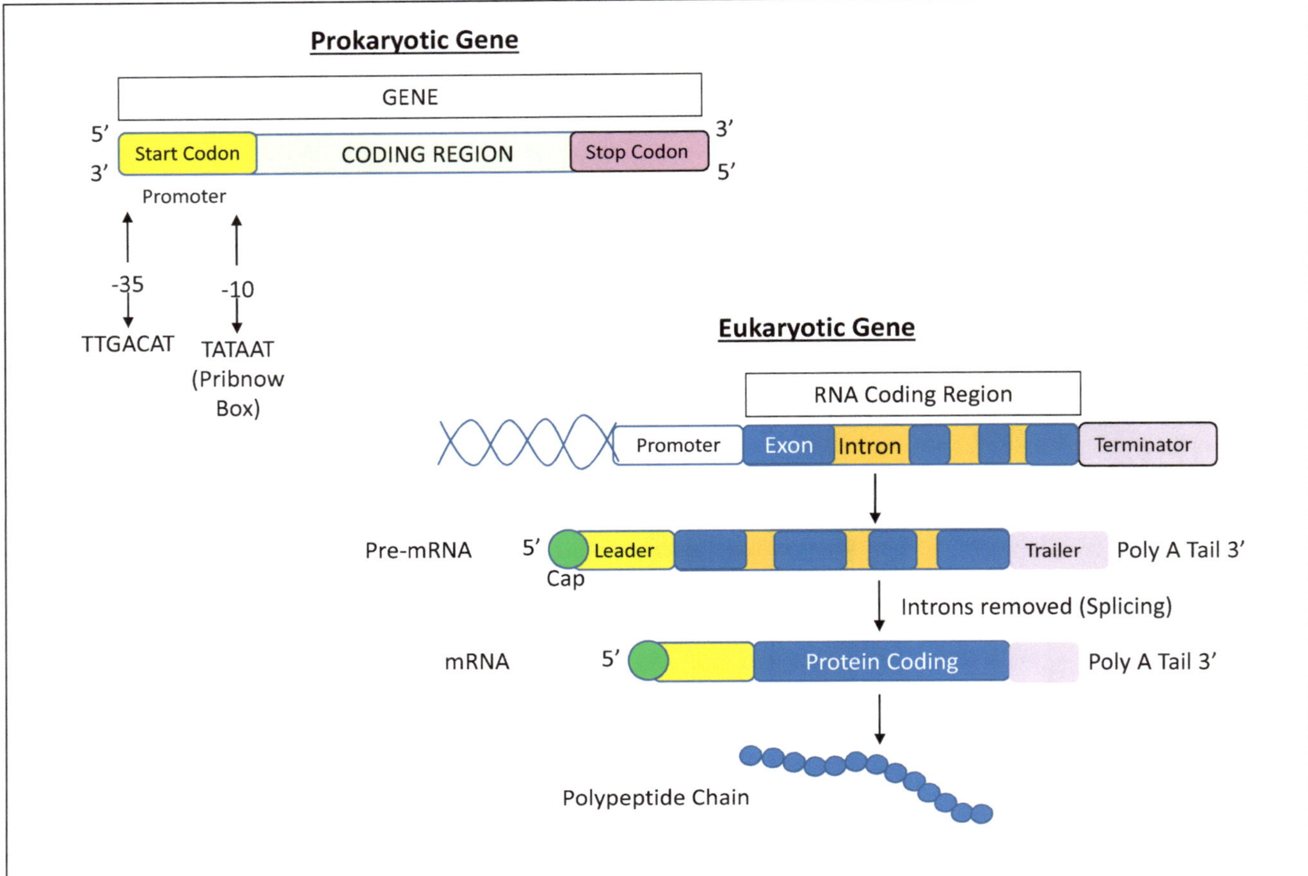

Fig. (12.6). Prokaryotic and Eukaryotic gene.

Requirement for Transcription	***Template*** Single strand of DNA	***Substrate*** 4 ribonucleoside triphosphate: ATP, GTP, CTP, UTP	***Direction*** After 1st NTP, subsequent Ntd are added to 3' of preceding nucleotide. RNA growth proceeds 5'→3'	***Enzyme*** RNA pol

Transcription Process	→DNA double helix partially unwind in the region of mRNA synthesis	**Three Steps**		
		Initiation	Elongation	Termination

→DNA is converted into a complementary RNA, catalyzed by RNA polymerase	→DNA template is unwound and rewound	→RNA polymerase complex assembles at promoter to initiate elongation of RNA	→Termination occurs when transcription complex responds to specific termination signals and disassembles
Four steps: *1.* ***Initiation****			

DNA unwinds and separates to form a small open complex involving template strand, RNA polymerase and RNA molecule: initiation complex	RNA polymerase binds to promoter of template strand: -Loosely binds 35 promoter sequence -Then tightly binds 10 promoter sequence	RNAP untwists DNA strand by about 17 bp to create a small bubble

2. ***Elongation***	RNA polymerase moves along template strand to synthesize mRNA
In *prokaryotes:*	RNA polymerase is a holoenzyme consisting of a number of subunits, including a sigma factor (transcription factor) that recognizes promoter RNA pol holoenzyme begins to synthesize RNA (in 5' →3' direction) by reading template strand (3' to 5') Sigma factor is released after transcript has 8-9 new RNA nucleotides RNA pol core enzyme continues to add NTPs to growing RNA strand, transcription bubble moves down stream at 30-50 ntd/sec across the gene.
In *eukaryotes*:	There are three RNA polymerases: I, II and III The process includes a proofreading mechanism.
3. ***Termination***	Prokaryotes and eukaryotes use identical mechanism of synthesizing RNA.
In *prokaryotes:*	Two ways in which transcription is terminated:

<table>
<tr><td></td><td colspan="3">In Rho-dependent termination: "Rho"(ρ) protein disrupts termination complex, Rho factor is ATPase

In Rho-independent termination: a loop forms at the end of RNA molecule, causing it to detach itself</td></tr>
<tr><td rowspan="4">In eukaryotes:</td><td colspan="3">It is more complicated, exact mechanism unknown</td></tr>
<tr><td>RNA pol I:</td><td colspan="2">Terminates transcription in factor-dependent manner</td></tr>
<tr><td>RNA pol III:</td><td colspan="2">Terminates by unknown mechanism

Terminal signals?</td></tr>
<tr><td colspan="3">Involves addition of adenine nucleotides at 3' of RNA transcript: polyadenylation</td></tr>
<tr><td>4. <u>Processing</u></td><td colspan="3">After transcription, RNA molecule is processed in a number of ways:

introns are removed and the exons are spliced together to form a mature mRNA molecule consisting of a single protein-coding sequence.

RNA synthesis involves normal base pairing rules, but base thymine is replaced with base uracil.</td></tr>
<tr><td>* Details of Initiation</td><td>Promoter Region</td><td><u>Start Codon</u>

AUG (Met)

TATA box for transcription factor.</td><td><u>Promoter Sequence</u>

Transcription initiation does not require primer.</td></tr>
<tr><td>Prokaryotic Promoter</td><td><u>Initiation Site</u>

Transcription for most genes always starts at same base (position): start point is usually a purine.</td><td><u>Pribnow Box</u>

9-18 bp upstream of start point termed: -10 sequence

Sequence similar or identical to: TATAAT</td><td><u>Another Promoter Component</u>

-35 sequence

Sequence similar as identical to: TTGACA</td></tr>
</table>

<table>
<tr><td>Eukaryotic Promoters</td><td>Initiation: Purine normally

Multiple RNA pol II sequence recognition element.</td><td colspan="2"><u>Promoter Sequence in Eukaryotes</u>

Each type of RNA pol uses different promoter

RNA pol I and II: similar to prokaryotic promoter

RNA pol III: promoter used by it is unique downstream of start point</td></tr>
<tr><td>Initiation site:

A purine</td><td>RNAP II promoter sequence:

Conserved: TATA box, CAAT box and GC box</td><td>RNAP III promoters:

SS RNA promoter:

Two sequence: 50-70 bp downstream, 80-90 bp down stream</td><td>tRNA promoter:

Made of two sequences:

Ist: between +8 and +30

2nd: +50 and +70</td></tr>
<tr><td colspan="4">Initiation Factor</td></tr>
<tr><td><u>In Prokaryotes</u></td><td colspan="3">• Only single factor: Sigma (σ) factor

• Function of σ factor: Enable RNA pol to recognize and tightly bind to promote sequences.

• Sigma factor facilitates opening or melting of DNA double helix, formation of phosphodiester bond between two bases by RNAP

• Elongation proceeds with formation of first phosphodiester bond and when 10 ntd are added, sigma factor dissociates

• Core enzyme (RNAP) continues further elongation of transcript</td></tr>
<tr><td>Rifampin:</td><td colspan="3">• Binds β-subunit of RNAP: antitubercular drug</td></tr>
<tr><td><u>In Eukaryotes</u></td><td colspan="3">Multiple factors are required:

RNAP II: required to initiate transcription from TATA box promoters.

IF: needed for initiation: important for control of numbered according to RNA pol number on lettered on order to discovery.</td></tr>
</table>

<table>
<tr><td></td><td colspan="2">Transcription factor II D (TFD II):
o Recognizes and bind to TATA box sequences independently of RNAP II:
o RNAP I & III require specific transcription factors, enhancers, response elements.
o TATA box-binding proteins (TBP): subunit of TFD II binds to TATA box DNA sequence
o TFD II has 8 subunits: TFD II associated proteins</td></tr>
<tr><td rowspan="2">Response Elements</td><td>Enhancers</td><td>Stimulate transcription rate located thousands of bp upstream or down stream</td></tr>
<tr><td>Silencers</td><td>Inhibit transcription</td></tr>
<tr><td colspan="3">Comparison</td></tr>
<tr><td></td><td>Eukaryotes</td><td>Prokaryotes</td></tr>
<tr><td>RNA Polymerase</td><td>3 types of RNA polymerase present</td><td>1 type of RNA polymerase present</td></tr>
<tr><td>Promoter Sequences</td><td>TATA box at -30*</td><td>Pribnow box at -10</td></tr>
<tr><td>Regulation</td><td>Complex: RNA polymerase requires several transcription factors to initiate transcription.</td><td>Simple: RNA polymerase only requires one additional protein (s factor).</td></tr>
<tr><td colspan="3">*(present 30 base pairs downstream from coding region)</td></tr>
</table>

TRANSLATION

<table>
<tr>
<td>Process where mature mRNA molecule is used as a template to assemble a series of amino acids to produce a polypeptide with a specific amino acid sequence according to the message's instruction.

Transcription and translation are divided in eukaryotes between nucleus (transcription) and cytoplasm (translation).

Requires energy.</td>
<td>Process

Three phases:

Initiation:

Ribosome binds at the 5' end of mRNA and the first amino acid attaches to its tRNA (start codon AUG).

Chain Elongation:

Ribosome adds one amino acid at a time to the growing poly-peptide chain using enzyme peptidyl transferase.

Termination:

Ribosomes release mRNA and polypeptide. Translation in terminated when the ribosomal complex reached one or more stop codons (UAA, UAG, UGA).</td>
</tr>
<tr>
<td colspan="2">Comparison: Prokaryotic Vs. Eukaryotic Translation

• In eukaryotes, translation does not occur simultaneously, transcription occurs in nucleus and translation occurs in cytoplasm

• In eukaryote, start codon AUG is translated to methionine.

• Eukaryotic mRNA encodes a single protein at a time, unlike bacterial mRNA which encodes many.

• Eukaryotic DNA contains introns which have to be spliced out of final mRNA transcript</td>
</tr>
</table>

Process ***Formation of Amino acyl-transfer RNA (tRNA):***	***Initiation:***
• *Activation*: Amino acids are activated and get attached to their corresponding tRNAs by amino acyl tRNA synthetase enzyme • Amino acid first reacts with ATP, forming amino acyl monophosphate (AMP) which forms an ester with 2' or 3'OH group of tRNA specific for that amino acid, producing amino acyl tRNA and AMP. • Once an amino acid is attached to tRNA, insertion of amino acid into a growing polypeptide chain depends on codon-anticodon interaction.	• mRNA attaches to 30S. • tRNA binds to mRNA where nucleotide code matches: This trigger binding of 50S to 30S. • tRNA binds to 50S. • First t-RNA binds to start codon in P site of ribosome. • AUG is start codon of every protein: AUG codes for amino acid methionine. • When second tRNA binds to A site, amino acid of 1st tRNA forms a peptide bond with amino acid of second tRNA.
50S Ribosome has Two Binding Sites: A, P sites	First tRNA attaches to A site, then moves to P site as other tRNA approaches 2nd tRNA binds to A site
Peptide bonds forms between amino acid of tRNA:	• 1st tRNA now gets detached from its amino acid and leaves ribosome • 2nd tRNA still has two amino acids attached.
Elongation:	• tRNA left now moves to P site. • Ribosomes ready to accept another tRNA and continue the process. • Each tRNA adds another amino acid to growing peptide chain (thus "elongation"). • Ribosome is moving along nucleotide triplets one by one.

Termination:	<ul><li>Ribosome reaches stop codon; peptide chain formation is fstopped.</li><li>Last tRNA leaves ribosome, leaving behind completed peptide chain.</li><li>Ribosome separates from mRNA.</li><li>Ribosome subunit also separate and remain in this way until mRNA comes along to restart the process.</li></ul>

Comparison of Protein Synthesis in Prokaryotes and Eukaryotes	**Identification Initiator Codon in Prokaryotes**
Although bacteria and eukaryotes carryout transcription and translation in similar ways, they do have differences in cellular machinery and processes:<ul><li>Different RNA polymerase and transcription factors</li><li>Differ in termination of transcription</li><li>Ribosomes are different</li><li>Prokaryotes can transcribe and translate the same gene simultaneously.</li></ul>	Involves binding of initiator tRNA (N-formylmethionyl tRNA) to initiator codon (first AUG) 30S subunit scans mRNA for a specific sequence: *Shine – Dalgarno (SD) sequence*: at upstream of initiator codon. 16 SRNA involved in recognition of S-D sequence

Details of Initiation	
Initiation in Prokaryotes	***Eukaryotic Initiation***
<ul><li>Initiation complex formed by protein initiation factors:<ul><li>IF-3 keeps ribosome subunits apart</li><li>IF-2 identifies and binds initiator tRNA</li></ul></li></ul>	<ul><li>Methionyl tRNA metabolic binds to small ribosomal subunit</li><li>5' cap of mRNA binds to small subunit</li><li>First AUG codon base pairs with anticodon on methionyl $tRNA^{Met}$.</li></ul>

▪ IF-2 binds GTP to bind tRNA ▪ IF-1, IF-2, IF-3 bind to 30S subunit to form initiation complex. o Once 50S subunit binds initiation complex, GTP hydrolyzed and initiator tRNA enters P site and IFs dissociate. o Prokaryotes do not contain 5' cap on their mRNA.	o No S-D sequence o CAP- binding protein (CBP) binds to 5'CAP structure of mRNA, forming *initiation complex* with initiation factor and 40S subunit. o Methionyl tRNA met is bound at P site A site is unoccupied. o Complex scans mRNA for 1st AUG near 5' end of mRNA. o eIF-2 transfers tRNA to P site and GTP hydrolyzed.
Sites P = peptidyl A = amino acyl	***Applied Aspect*** Tetracycline: bind to 30S ribosomal subunit, blocks access of amino acyl tRNA to mRNA ribosomal complex.
Details of Elongation	***Applied***
Elongation	Puromycin: analog of amino acyl tRNA inhibits translation by chain terminator. • EF-Tu binds to amino acyl t-RNA and delivers it to A site of ribosome • Ef-Tu binds GTP causing conformational change, allowing it to bind to amino acyl tRNA. • EF-Tu-tRNA complex enters ribosome and positions new tRNA at A site
• Correct amino acyl t-RNA occupy position at acceptor site • Formation of peptide bond between peptidyl t-RNA at P site with amino acyl tRNA at A site. • Translocation of peptidyl tRNA to P site • Shifting mRNA by codon relative to ribosome	

<table>
<tr><td>

Binding of Amino acyl-tRNA to A Site

- mRNA codon at A site determines which amino acyl tRNA will bind
- Codon and anticodon bind by base pairing.
- Internal methionine residues in polypeptide chain are added in response to AUG codon.

They are carried by $tRNA_m^{Met}$: a second tRNA specific for methionine.

- Elongation factor and hydrolysis of GTP are required for binding.

</td><td>

Formation of Peptide Bond

- Peptide bond forms between amino group of amino acyl tRNA at A site and carbamoyl of amino acyl group attached to tRNA at P site by peptidyl transferase.
- tRNA at P site now does not contain an amino acid termed: UNCHARGED
- Growing polypeptide chain is attached to tRNA in A site

</td></tr>
<tr><td>

Translocation of Peptidyl tRNA

- If moves from A to P site (along with attached mRNA) and uncharged tRNA released from ribosome.
- Elongation factor (EF-2/EF-G) and GTP hydrolysis required for translocation.
- Next codon in mRNA is now in A site.
- Elongation and translocation steps are repeated till termination codon moves into A site.

</td><td>

Applied Aspect

Diphtheria toxin causes ADP-ribosylation of EF-2 inhibiting translation in eukaryotes.

- If anticodon matches the codon, GTP hydrolyzed, EF-Tu releases tRNA and exits ribosome.

</td></tr>
</table>

Analogous Elongation Factors in Prokaryotes and Eukaryotes

Prokaryote	***Eukaryotic***	
EF-Tu	EF-1a	- Docks tRNA in A site
EF-Ts	EF-1b	- Recycles EF-Tu

EF-G	EF-2	- Involved in translocation
Details of Termination:		
Termination	When stop codon (UGA, UAG, UAA) occupies A site, release factors cause release of polypeptide chain from ribosomes and ribosomal subunits dissociate from mRNA.	
Polysomes	More than one ribosome can be attached to a single tRNA at a time in prokaryotes and this complex of mRNA with multiple ribosomes is POLYSOME.	

Co-and Post Translational Covalent Modifications	
Process	***Time of Occurrence***
Disulfide bond formation Glycosylation	Protein folding as protein pass through ER and golgi
Proteolysis	Cleavage of peptide bonds to remodel proteins and activate them, *viz.* activation of Proinsulin, Trypsinogen, Prothrombin
Phosphorylation	Addition of PO_4^{3-} converts uncharged protein to negatively charged hydrophilic protein, Kinases phosphorylate & phosphatases dephosphorylate proteins. *E.g.*: regulation of p53, Na^+/K^+-ATPase
γ-carboxylation	Addition of carboxylate group by vitamin K dependent carboxylase produces Ca^{+2} binding sites
Prenylation	Addition of farnesyl or geranyl groups to certain membrane associated proteins

MUTATION

<table>
<tr><th colspan="2">Fact File</th><th>Causes</th></tr>
<tr><td colspan="2">Permanent changes in a DNA sequence

Creates new alleles

Depending on location and type of mutation can be beneficial, neutral or detrimental</td><td>• Occurs spontaneously, often during replication
• Replication errors: point mutation, insertion, deletions
• Some form during meiosis
• Errors during recombinant events
• Chemical mutagens
• Irradiation</td></tr>
<tr><td rowspan="4">Types
(Fig. 12.7)</td><td>Base Substitution

Transition, Transversion</td><td rowspan="4">Mis Sense, Non-sense or Silent Mutation

<u>Mis Sense</u>

Result in change in amino acid in protein product of gene

TGT → TGG

Cys → Trp

E.g. sickle cell anaemia

<u>Nonsense</u>

Result in conversion of amino acid codon to stop codon

TGT → TGA</td></tr>
<tr><td>Deletion:

Deletion of one or more base pairs:

Frame shift mutation, producing non-functional protein</td></tr>
<tr><td>Insertion:

Insertion of one or more base pairs</td></tr>
<tr><td>Somatic Mutation

• Occurs in somatic cells</td></tr>
</table>

	• All daughter cells from mutated cells will contain the mutation	Cys $\rightarrow$ Stop *E.g.* Thalassemia ***Silent Mutation*** Result in formation of a codon synonym No change in amino acid sequence of gene product TGT $\rightarrow$ TGC Cys $\rightarrow$ Cys
	Germ Line Mutation • Heritable mutations • Germ cell mutations are perpetuated in gametes and passed on to offspring.	

Importance of Mutations
 	
• Create new alleles
  ○ Allow bacteria to become antibiotic resistant allows viruses to infect new hosts
• Induced mutations let scientists study gene function:
  ○ Genes with induced mutations differ in function from normal gene
  ○ Allows scientists to determine how a gene works:
    - Site- directed mutagenesis
		Deamination
		
		Deamination of cytosine converts it into uracil
		
		Deamination of adenine converts into hypoxanthine
		
		Deamination of guanine converts into thymine
		
		5-methylcytosine into thymine

Mutation by Replication Errors Replication error: Main source of mutations Occur at frequency of 10^9-10^{11}		***Mutation by UV Light*** • UV light may cause two adjacent pyrimidine residues (cytosine or thymine) to form a dimer. • p53 can detect this dimer in normal cells and then triggers repairing process. • It p53 in mutated, pyrimidine dimmer may lead to mutation *E.g.*, Cytosine dimer could cause adenine to be incorporated into new strand. Subsequent DNA

Since cell division requires synthesis of 6 x 10^9 nucleotides, mutation rate is about one per cell division.

- Commonly observed replication errors: replication slippage
- Occurs at repetitive sequences when new strand mispairs with template strand *e.g.* microsatellite polymorphism.
- If mutation occurs in a coding region, its produces abnormal proteins leading to diseases *e.g.* Huntington's disease

replication could cause adenine to be incorporated into new strand.

Chemical Mutagens

Chemical agents that may cause mutation are called chemical mutagens. Most of them are also carcinogens.

Base Analogs	5-bromouracil 2 aminopurine Acridine
Alkalylating Agents	Dichyloroethyl sulphide (sulfur mustard) Dichyloroethyl methylamine (nitrogen mustard) Ethylanethane sulfonate (EMS)
Deaminating Agents	Nitrous acid
Miscellaneous	Hydroxylurea Free radicals

Consequences of Mutation	***→ Loss of Functions***	***→ Gain of Function***
Changes in base sequence of DNA can be lethal and	- Complete loss of functions:	- Change in gene product such that it gains a new

inheritable ***Effect on function:*** →	- amorphic mutation - Occurs in recessive phenotype.	and abnormal function: neomorphic - Occurs in dominant phenotype

Dominant Negative Mutation Altered gene product causing altered molecular function *E.g.* Marfan's syndrome: Defective fibrillin gene product	*Xeroderma pigmentosum (XP):* - Hypersensitive to sunlight - UV induced pyrimidine dimers - Increased incidence of skin cancer, premature aging.

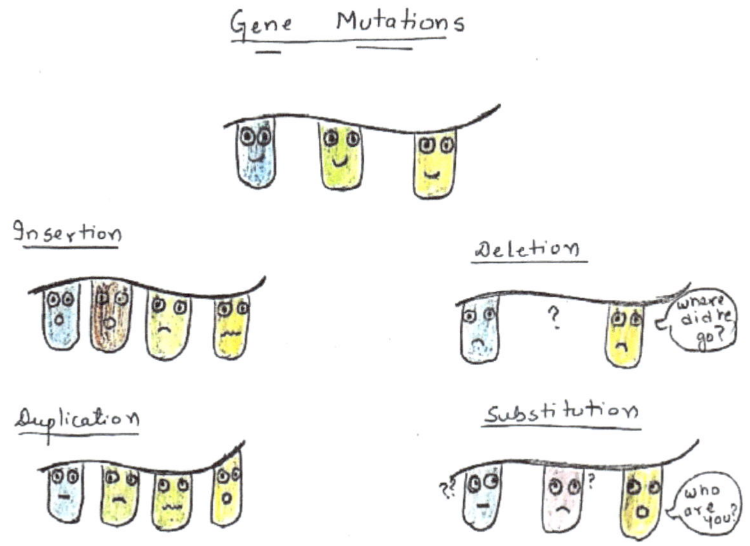

Fig (12.7). Gene mutations.

REGULATION OF GENE EXPRESSION

Gene Expression	Orderly use of genomic information during development and for response to changes in both internal as well as external environment.

Fact File About~1.4% of the human genome (3.2 billion base pairs) encodes the proteins. There are around 23,000 protein-encoding genes in mammalian genomes, and 42% of them are referred as *operon* as their functions are unknown function. *About 60% of human genome transcribes RNA:* > 4000 genes encode rRNA, tRNA, snRNA and small RNAs; In another 10,000 loci, transcription of noncoding RNAs takes place	***Prokaryotic Operons*** They include genes related to a specific metabolic function. ***In Eukaryotes*** Genes for histones and globins occur in clusters. ***A Significant Portion of Prokaryotic and Eukaryotic Genomes is Never Transcribed:*** ***<u>In Prokaryotes</u>:*** They are predominant in promoter and operator sequences ***<u>In Eukaryotes</u>:*** They include promoters and other regulatory regions and bulk of them are repetitive DNA with unknown function.
***ncRNAs*:** Have no known function and many of them are involved in RNA interference (a form of posttranscriptional gene silencing). Throughout the genome, genes are not distributed randomly, and some genes are organized in both prokaryotic and eukaryotic genomes.	Also, moderately repetitive sequences ($< 10^6$ copies per haploid genome) occur in 100 to several thousand bp segments and most of them are retrotransposons. In 42% of human genome, three types of retrotransposons are present, namely: 6–8 kb long interspersed elements (LINEs); 100-400 bp long short interspersed elements (SINEs); and long terminal repeats (LTRs).

<table>
<tr><td colspan="3">Regulation of Prokaryotic Gene Expression</td><td colspan="2">Regulation of prokaryotic gene expression occurs mainly at transcriptional level.</td></tr>
<tr><td colspan="2">• Lac operon</td><td colspan="2">• Histidine operon</td><td>• Tryptophan operon</td></tr>
<tr><td colspan="2">Operon</td><td colspan="3"><u>Two Types</u></td></tr>
<tr><td colspan="2">A coordinated set of genes, all of which are regulated as a single unit</td><td colspan="2"><u>Inducible</u> -Operon is turned on by substrate</td><td><u>Repressible</u> -Operon is turned off by the product synthesized.</td></tr>
<tr><td>Lactose Operon (Fig. 12.9)
3 segments</td><td colspan="2"><u>Regulator</u>

<u>Control Locus</u>

<u>Structural Locus</u></td><td colspan="2">gene the code for repressor

composed of promoter and operator

Made of 3 genes, each coding for an enzyme required to catabolic lactose:

β-Galactosidase – hydrolyze lactose

Permease – brings lactose across cell membrane

β-galactosidase transacetylase uncertain function</td></tr>
<tr><td>Lac Operon</td><td colspan="2"><u>Normally off</u>

In absence of lactose, repressor binds operator locus and blocks transcription of downstream structural gene.</td><td colspan="2"><u>Lactose Turns Operon On</u>

• Binding of lactose to repressor protein changes its shape and causes it to fall off the operator.

RNA pol can bind to promoter, structural genes are transcribes

• Lactose inhibits repressor protein

• cAMP dependent activator protein binds CAP site

• Glucose lowers cAMP, preventing activation

• Genes are expressed when lactose is present and glucose is absent.</td></tr>
</table>

Regulation of Eukaryotic Gene Expression	
• Chromatin modifying activities: ○ Histone acetylase (favor gene expression) ○ Histone deacetylase (favor inactive chromatin)	• Response elements bind to DNA: ○ Upstream promoter elements ○ Enhancer response element
• DNA methylating enzymes (favour inactive chromatin)	• Scaffolding protein (condensing regions of chromatin)- favor inactive chromatin • Activator proteins
• Repressors bind silencer elements • Activators (transcription factor): ○ Specific transcription factors: Steroid receptors (Zn finger) - cAMP – dependent activator protein, CREB (leucine zipper) - Homeodomain proteins: pHOX, pPAX (helix-turn-helix)	***Zinc Finger Structure*** Amino acid holds Zn^{+2} ions Include 2 residues of cysteine (C) with 2 histidine (H) residues (Fig. **12.8**). ***Leucine Zipper*** Leucine residues as spaced exactly every 7 amino acids Interaction of leucine side chains helps hold the 2 helices together.

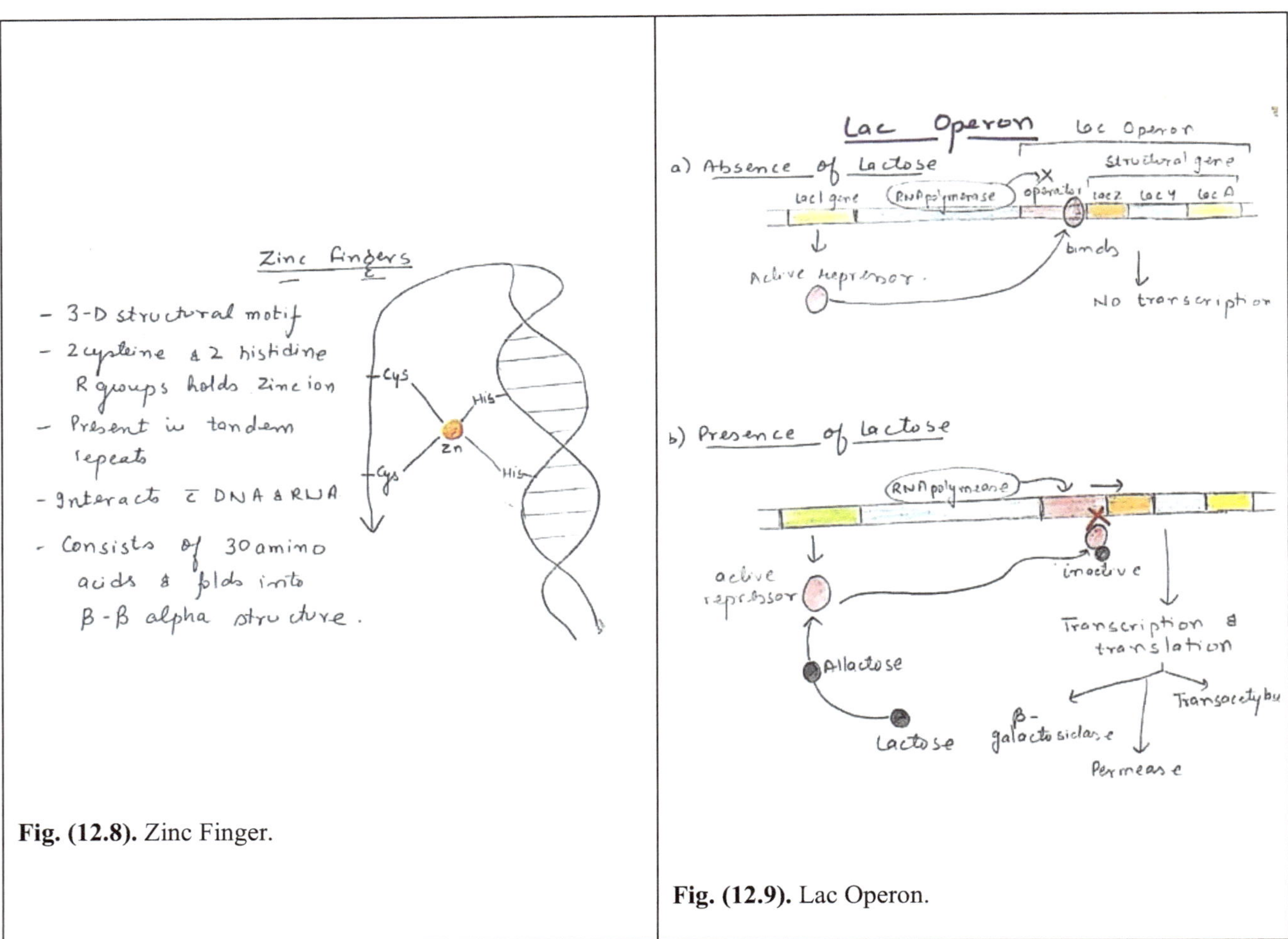

Fig. (12.8). Zinc Finger.

Fig. (12.9). Lac Operon.

BIOCHEMISTRY OF CANCER

Oncogenesis	Basic Properties of a Cancer Cell
Cells proceed through multiple rounds of division during growth and maintenance of an organism. Cells may be subjected to various insults *e.g.*, chemical, radiant energy or viruses,	Most important is loss of growth control. They ignore inhibitory signals and continue to grow even in the absence of stimulatory growth signals.

If the normal DNA repair mechanisms become overwhelmed, they may cause chemical changes, resulting in mutations. Numerous mechanisms exist to create genetic changes that result in uncontrolled cell growth.	They s do not undergo aging and continue to divide indefinitely due to presence of telomerase They fail to elicit apoptotic response. They are in many cases depend on anaerobic metabolic pathways *e.g.* glycolysis to a much greater degree than their normal counter parts. They have tendency to spread to distant sites within the body.

Cancer Development	A multistep process Three main phases: initiation, promotion and progression.
Initiation	First step in oncogenesis. Involves one or more stable cellular changes that arise either spontaneously or are induced by exposure to a carcinogen. Cellular genome undergoes mutations, creating a potential for neoplastic development. Transformed cell continues to divide with transformed karyotype and further mutations, till a malignant lesion is manifested.
Promotion	Transformed (initiated) cell may remain harmless or may get undergo further proliferation and later on undergo neoplastic transformation following prolonged exposures to promoting stimuli. With tumor progression, cells lose their adherence property, get detach from tumor mass and may invade the neighboring tissues or may finally enter circulation to produce distant metastasis

<table>
<tr><th colspan="3">Cancer Genes</th></tr>
<tr><td>Protooncogenes</td><td colspan="2">They are the normal cellular proteins that work to regulate normal growth and development.

If they are mutated or mis expressed, then they become oncogenes and lead to aberrant cell cycle control.

Aberrant expression of oncogene causes entry of cell into cell cycle with abnormal cell growth and gain of function. They function in growth or signaling pathways.</td></tr>
<tr><td>Activation of Oncogenes</td><td colspan="2">• Mutation of gene: that alters the property of gene product
• Duplication of gene causing gene amplification produces excess of encoded protein.
• Chromosomal rearrangement: cause/ alter expression of gene.</td></tr>
<tr><td>Genes Involved in Carcinogenesis</td><td colspan="2">They include tumor suppressor gene, oncogenes.

They produce products involved in control of cell cycle, intercellular adhesion and DNA repair.</td></tr>
<tr><td>Tumor Suppressor Genes</td><td colspan="2">They encode proteins that restrain the cell growth and prevent cells from becoming malignant. Oncogenes: encode for the proteins that promote loss of growth control and malignancy.</td></tr>
<tr><td>Oncogenes</td><td>They arise from protooncogenes genes that encode proteins having a role in cell's normal activities.

Most tumors contain alterations in both tumor suppressor genes and oncogenes.

They are derived from protooncogenes:</td><td>Oncogenes encode proteins that promote loss of growth control and conversion of a cell to a malignant one.

Most oncogenes play role in control of cell growth and division i.e., they act as accelerator of cell proliferation. Different oncogenes get activated in different types of tumors.</td></tr>
</table>

	Genes that encode proteins having a function in normal cell.	Oncogenes act dominantly; single copy can cause the altered expression.
Protooncogene	***Neoplasm***	***Lesion***
ABL	Chronic myelogenous leukemia	Translocation
BCL2	B-cell lymphoma	Translocation
ERB B	Squamous cell carcinoma	Amplification
NEU/HER 2	Adenocarcinoma Breast, ovary, stomach	Amplification
MYC	Burkitt's lymphoma Carcinoma lung, breast, cervix	Translocation Amplification

Ways of Activating (proto-) Oncogenes

	Mechanism	*Oncogene*	*Tumor*
1.	Amplification	ERB B2(HER 2)	Breast, ovary, lung, gastric, colon cancer
		NYCN	Neuroblastoma
2.	Point mutation	HRAS	Bladder, lung, colon cancer
		NIT	GIT tumor, mastocytosis
3.	Chromosomal rearrangement	BCR=ABL 1	Chronic myeloid leukemia
4.	Transduction to a region of transcriptionally active chromatin	MYC	Burkitt's lymphoma

Tumor Suppressor Genes	Tumor suppressor genes are cellular proteins whose activity if reduced, results in uncontrolled cell growth.
Common tumor suppressor genes: RB TP53 APC BRCA 1 and BRCA 2	As long as a cell has its full complement of tumor- suppressor genes, it is protected against the effects of oncogenes. Transformation of a normal cell to a cancer cell is accompanied by loss of function of one or more tumor suppressor genes. Tumor suppressor genes are often silenced epigenetically by methylation of promoter containing CpG islands.
APC Gene	Adenomatous polyposis coli gene: promotes degradation of β catenin which normally translocates to nucleus to induce cellular proliferation
FAP	Familial adenomatous polyposis: results from mutations of APC gene.
Rb Gene	Negative regulator of nuclear transcription, loss of Rb function predisposes to retinoblastoma, osteosarcoma
p53 Gene	Gate keeper of cellular proliferation.

RB Gene	
Responsible for retinoblastoma – rare, familial retina tumor. Negative regulator of cell cycle. RB gene codes a protein pRb, which is involved is regulatory passage of a cell from G, to S in cell cycle	Dephosphorylated form of pRb interacts with certain transcription factor (E2F) preventing them from activating genes required for S phase activities Once pRb phosphorylated, it releases its bound transcription factor, that can activate gene expression to initiate S-phase
p53: Guardian of Genome Induced when DNA is damaged	*P53* can direct this cell towards apoptosis. *P53* inhibits CDK- and cyclin- mediated phosphorylation of Rb (required for transition to S-phase)

Has growth inhibitory function Product of TP53 tumor suppressor gene Acts as transcription factor that activates expression of p21 protein that inhibits cyclin – dependent kinase that moves a cell through cell cycle. *p21* protein inhibits cyclin-dependent kinase that normally drives a cell through G_1 check point Damage to DNA triggers phosphorylation and stabilization of p53, leading to arrest of cell cycle, until the damage can be repaired. Failure to repair DNA damage leads to production of abnormal/malignant cells via activation of expression of BAX gene (encodes protein Bax that initiates apoptosis).	Because of its ability to trigger apoptosis, p53 plays a pivotal role in treatment of cancer by radiation and chemotherapy. If a cancer cell loses p53 function, it cannot be directed into apoptosis and become highly resistant to further treatment (*E.g.* colon cancer, prostate cancer, pancreatic cancer responds poorly to radiation and chemotherapy). ***Mutation of p53*** Common molecular alteration in cancer Account for 50% of mutations in tumors Germline mutation of p53 causes Li Fraumeni syndrome Sporadic mutation can cause sarcoma, breast cancer, tumors of central nervous system and leukemia.

Apoptosis Fact File	***Definition***	Programmed destruction of cell
-Daily $10^{10} – 10^{11}$ cells in human body die -During embryonic development extra neurons (that fail to find their target tissues), surplus T-cells are destroyed by apoptosis.	***Characterized by***	Decrease in cell volume Mitochondrial destabilization Chromatin condensation Nuclear fragmentation Cellular dispersion into fragmented apoptotic bodies.
-Cells with irreparable genomic damage are destroyed by apoptosis normally. -Apoptosis is involved in neurodegenerative disorders (Alzheimer, Parkinson's, Huntington's)		

Programmed Cell Death (PCD)	Very large numbers of cells of a multicellular organism are deliberately and naturally selected to die throughout its existence. A variety of different types of PCD are known, of these apoptosis (type I PCD) has been extensively studied and characterized by specific changes in cell structures.
Apoptosis Results in Cell Death ***Mechanism***	• Withdrawal of growth factor. • Proapoptotic cytokines, tumor necrosis factor (TNF), fas ligand provide signals for apoptosis to stimulate pro apoptotic enzymes: CASPASES • Activation of proapoptotic gene: Bax by tumor suppressor gene, p_{53} in case of detection of DNA mutation at G_1/S checkpoint.
Death by apoptosis is a neat, orderly process, characterized by shrinkage of volume of cell, its nucleus, and loss of adhesion to neighboring cells, formation of blebs on cell surface, fragmentation of chromatin and phagocytosis of this dead cell.	
Final Events of Apoptosis	• Cyt c in outer mitochondrial membrane is critical regulator of apoptosis • Normally bcl-2 (antiapoptotic) gene inhibit translocation of cyt c out of mitochondria • Cyt c exists via bax channel protein • Balance between bcl-2 and bax determine fate of cell, if bax predominates, cyt c is liberated to associate with proapoptotic protease activating factor (Apaf-1) • Apaf-1 activates cascade of proteolytic events via activation of caspases.
Caspases	• C: cysteine
Aspase	• aspartic acid protease activity • Exist in cytoplasm as zymogen

<table>
<tr><td colspan="4"></td><td colspan="4">

- On stimulation begin apoptotic cascade.
- They degrade intracellular proteins and activate DNases with resultant DNA fragmentation.

</td></tr>
<tr><td colspan="8">Targets of Caspases</td></tr>
<tr><td colspan="2">Protein kinases: focal adhesion kinases (FAK), PKB, PKC, Raf1.</td><td colspan="2">Lamins in inner nuclear membrane</td><td colspan="2">Cytoskeleton proteins: intermediate filaments, action, tubulin, gelvolin</td><td colspan="2">Endonuclease termed caspase activated DNase (CAD)</td></tr>
<tr><td colspan="8">Stimuli for Apoptosis</td></tr>
<tr><td colspan="2"><u>Apoptosis Can Be Triggered Both By</u></td><td colspan="3">Internal stimuli (e.g. abnormality in DNA):

Intrinsic pathway.</td><td colspan="3">External stimuli (e.g. cytokine):

Extrinsic pathways</td></tr>
<tr><td colspan="2">Extrinsic Pathway of Apoptosis</td><td colspan="3">TNF produced by certain cell of immune system in response to:

Ionizing radiation

Elevated temperature

Viral infection

Toxic chemical agents (e.g. cancer chemotherapy drugs)</td><td colspan="3">TNF evokes its response by binding to its receptor TNFR 1.</td></tr>
<tr><td colspan="4"><u>Sequence of Events in Extrinsic Pathway</u>

TNF
+
THF – Receptor
(TNF-R1)
↓ Binds
TRADD, FADD
Procaspase B
↓</td><td colspan="4">TNFR 1:

- Transmembrane receptor
- Present as trimer
- Cytoplasmic domain contains death domain of 70 amino acid

</td></tr>
</table>

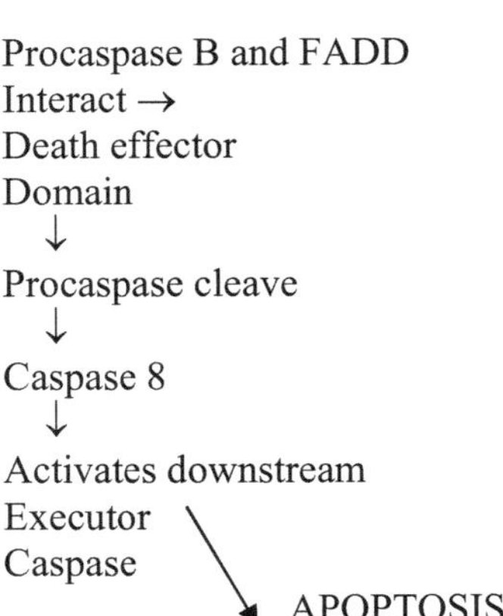

- Binding of TNF to TNFR 1 causes conformational change in its death domain and binds to adaptor proteins TRADD, FADD.
- Two procaspase-8 bind to death domain. They cleave one another to generate active caspase-8.
- Caspase-8 is **initiator** caspase and it activates downstream or **executor** caspases that carry out controlled self-destruction of cell

Intrinsic Pathway of Apoptosis	***Stimuli***	***BCL2***
	Irreparable genetic damage Lack of oxygen High cytosolic Ca^{+2} Severe oxidative stress.	Proapoptotic family of protein ***Two Groups*** *Proapoptotic:* Bad, Bax *Antiapoptotic:* Balx$_L$, Bcl-w, Bcl-2

Sequence of Events in Intrinsic Pathway

Internal cellular damage

↓ Activates

Bcl 2 (*e.g.* Bad, Bax)

↓

Base inserted into outer mitochondrial membrane

↓ Bax forms pores

Releases cyt c in cytosol

- Following intrinsic stimuli proapoptotic base translocates from cytosol to outer mitochondrial membrane and increases its permeability, releasing cyt c from it.
- Bax (and/or Bak) form channel within mitochondrial membrane. Activated along with raised cytosolic Ca^{+2} (coming from ER).
- Antiapoptotic protein Bcl-2 can inhibit release of cyt C directly or indirectly.

Apaf-1 and procaspose-9 bind to cyt c ↓ Apoptosome Activated initiator Caspase-9 complex → Procaspase 9 → caspase 9 → → → APOPTOSIS	• In cytosol, cyt C forms a complex with several procaspase 9 molecules called *apoptosome.* • Procaspase get activated on joining to apoptosome complex to form caspase- 9 • Caspase 9 is initiator caspase that activates downstream executioner caspases to bring about apoptosis.
Remember Extrinsic (receptor-mediated) and intrinsic (mitochondria-mediated) pathways ultimately converge on same executioner caspases that cleave same cellular targets. Entire apoptotic program can be executed in less than an hour. Apoptotic cell death occurs without spilling cellular content into extracellular environment because cellular debris can trigger inflammation.	**Sequence of Events Following Execution of Apoptotic Program** • Cells lose contact with neighbors and shrink • Cells disintegrate into apoptotic bodies • Apoptotic bodies recognized by presence of phosphatidylserine (PS) on their surface • Normally PS is on inner leaflet of plasma membrane, during apoptosis, phospholipids scramblase moves PS to outer leaflet of plasma membrane. • This (eat me signal) is recognized by macrophages.
Thus, fate of a cell: whether survival or death- depends on a delicate balance between proapoptotic and antiapoptotic signals.	

QUESTIONS

1. Define:

   a. Chargaff's rule

b. Coding strand

c. Template strand

d. End replication problem

e. Splicing

f. Ribozymes

g. Topoisomerases

h. Genetic code

i. Oncogenes

j. Caspases

2. Describe:

a. Structure of DNA

b. Histone modifications

c. Types of DNA polymerase

d. DNA replication

e. Telomerase

f. Types of RNA

g. Structure of t-RNA

h. DNA repair mechanisms

i. Post transcriptional modifications

j. Significance of mRNA capping and poly (A) tail

k. Inhibitors of protein synthesis

l. Posttranslational modification

m. Phases of cell cycle and its checkpoints

n. DNA binding proteins

3. Differentiate between:

a. A-DNA, B-DNA and Z-DNA

b. Denaturation and annealing of DNA

c. Topoisomerase I and topoisomerase II

d. RNA and DNA

e. DNA replication and transcription

f. Eukaryotic and prokaryotic promoters

g. Eukaryotic and prokaryotic RNA polymerase

h. Eukaryotic and prokaryotic translation

i. Normal and cancer cell

j. Tumor suppressor genes and oncogenes.

k. Apoptosis and cancer

4. What are the similarities of DNA and RNA polymerase?

5. Make a list of:

a. Drugs that inhibit replication with their mechanism of action

b. Disease associated with defective DNA repair

6. Explain the modified bases and their significance.

7. What are mutations? What are their types and consequences?

8. Write a brief note on eukaryotic gene regulation

9. Make a list of proto-oncogenes and cancer caused by them

10. Write a short note on significance of p53.

11. Describe Apoptosis, its biological significance and regulation.

12. Write a brief note on apoptosis and cancer

BIBLIOGRAPHY

Bruce Alberts, Alexander Johnson, Julian Lewis, Martin Raff, Keith Roberts, and Peter Walter. Molecular Biology of the Cell. 4th edition. New York: Garland Science; 2002.

Denise R Ferrier. Lippincott illustrated reviews: biochemistry. 7th Edition. Philadelphia Wolters Kluwer; 2017

Donald Voet, Judith G Voet, Charlotte W Pratt. Fundamentals of Biochemistry. 5th Edition. New York: Wiley; 2016.

Geoffrey L Zubay, Dennis E Vance. Principles of biochemistry. Dubuque, Iowa: William C. Brown; 1995.

Geoffrey M. Cooper & Robert E. Hausman. The cell: A molecular approach. 7th Edition. Oxford University Press; 2019.

Jeremy M Berg, Gregory J Jr Gatto, Lubert Stryer, John L Tymoczko. Biochemistry. 9th Edition. New York: Macmillan International Higher Education: WH Freeman; 2019.

Lehninger A, Nelson D, Cox M. Lehninger principles of biochemistry. New York: Worth Publishers; 2000.

Michael A Lieberman, Rick E Ricer. Biochemistry, molecular biology, and genetics. 7th Edition. Philadelphia, Pa Wolters Kluwer; 2020.

Victor W Rodwell, David A Bender, Kathleen M Botham, Peter J Kennelly, P Anthony Weil. Harper's illustrated biochemistry. 31st edition. New York: Mcgraw-Hill Education; 2018.

CHAPTER 13

Genetic Engineering

LEARNING OBJECTIVES:	Keywords:
• Describe the tools, techniques and applications of genetic engineering. • Explain the procedures and applications of PCR and cloning.	Cloning, Electrophoresis, PCR, Recombinant DNA Technology, Taq polymerase, Vectors.

MOLECULAR BIOLOGY TECHNIQUES: APPLICATIONS

Recombinant DNA Technology ***Three Basic Approaches***	***Fact File*** A gene might be 1/1,000,000 of genome	
PCR (Polymerase Chain Reaction)	Makes any copies of a specific region of DNA	
Cell-Based Molecular Cloning	Create and isolate a bacterial strain that replicates a copy of the gene of interest	
Hybridization	Make DNA single-stranded. Allow double-stands to reform using a labeled version of the selected gene to make it easy to detect.	
PCR (Polymerase Chain Reaction) (Fig 13.1)	Based on DNA polymerase property of creating a second strand of DNA	
Requirements	Template DNA	Two primers to flank the region to be amplified

Simmi Kharb

Primers	Short (18-30 bases) DNA oligomers complementary to the ends of the region being amplified	
DNA Polymerase	Adds new bases to 3' ends to primer to create a new second strand	
DNA polymerase from *Thermus aquatics*: ***Taq polymerase →***	A bacterium that lives in nearly boiling water in Yellow stone Natural Park Hot springs	Can with strand temperature cycle of PCR that world till DNA pol from *E. coli.*
Uses of PCR	Diagnosis of diseases: - infectious, cancer Forensic medicine Antenatal diagnosis Prenatal sex determination. Paternity contesting	Construction of useful organisms: - Bacteria of biological waste handling, marine spills, nitrogen fixation. Detection of mutations Preimplantation diagnosis of genetic diseases. Anthropology
Advantages of PCR	Rapid Sensitive	Robust Works even with partly degraded DNA
Disadvantages	Only a short region (up to 2kbp) can be amplified.	A limited amount of product is made.
Problems in PCR	Contamination, impurities, improper sample	
PCR Cycle	Based on the cycle of three steps that occur at different temperatures	

<table>
<tr><td></td><td colspan="3"><ul><li>Each cycle doubles the number of DNA molecules: 25-35 cycles produce enough DNA to be visualized on an electrophoresis gel.</li><li>Each step takes one minute to complete</li></ul></td></tr>
<tr><td>Steps</td><td><u>Denaturation</u>

Makes DNA single-stranded by heating to 94°C.</td><td><u>Annealing</u>

Hybridize primers to single strands

Temperature around 50°C</td><td><u>Extension</u>

Build second strand with DNA pol and dNTP: 72°C.</td></tr>
<tr><td>Electrophoresis</td><td colspan="3"><ul><li>Separation of charged molecules in the electric field</li><li>Nucleic acid have one charged phosphate (negative charged) per nucleotide</li><li>Separation based on length: longer molecule moves slower.</li><li>Done in a gel matrix to stabilize: agarose or acrylamide</li><li>Average run: 100 volts across 10cm gel, run for 2 hours</li><li>Stain with ethidium bromide: intercalates with DNA bases and fluoresces orange.</li><li>Run alongside standards of known sizes to get the length.</li></ul></td></tr>
<tr><td colspan="4">Cloning</td></tr>
<tr><td>Sources of DNA</td><td colspan="3">Genomic DNA: Whole-genome is cut into small pieces and cloned.</td></tr>
<tr><td>Methods</td><td colspan="3">Random shear by

- Partial digestion to generate recognition site at every 256bp by restriction enzyme</td></tr>
</table>

cDNA	• DNA copy of mRNA, with reverse transcriptase.
Synthetic DNA	• Synthesized *de novo* or by PCR.
Cell-based Molecular Cloning	
Restriction Enzymes	• Cut DNA at specific sequences *e.g.* EcoR I • Eco RI cut at GAATTC • Used by bacteria to destroy invading DNA their own DNA has been modified at corresponding sequences by a methylase
Plasmid	Independently replicating DNA circles (only circles replicate in bacteria) • Foreign DNA can be inserted into a plasmid and replicated • Plasmids for cloning carry drug resistance genes to help in the selection.
DNA Ligase	Attaches two pieces of DNA together
Transformation	DNA manipulated *in vitro* can be put back into the living cells by a simple process. Transformed DNA replicates and expresses its genes.
Cloning Vectors	Plasmid for up to 5kb Phage lambda (x): upto 50kb BAC (bacterial artificial chromosome): 300 kb YAC (yeast artificial chromosome): 200kb
Expression Vectors	Make DNA and protein from inserted DNA.
Basic Cloning Process	• Plasmid is cut open with a restriction enzyme that leaves on overhang: sticky end.

	• Foreign DNA is cut with the same enzyme • Two DNAs are mixed • Sticky ends anneal together • DNA ligase joins them into one recombinant molecule • Recombinant plasmids transformed into E. coli using heat, calcium chloride. • Cells carrying the plasmid are selected by adding on antibiotic: plasmid carries a gene for antibiotic resistance
Hybridization	DNA when made single stranded by melting, pairs up with another DNA or RNA with complementary sequence. If one of the DNA molecules is labeled, it can detected by hybridization.
Applications of Hybridization	Southern blot to detect DNA Northern blot to detect RNA *In situ* hybridization for probing a tissue Colony hybridization for detection of clones. Microarray

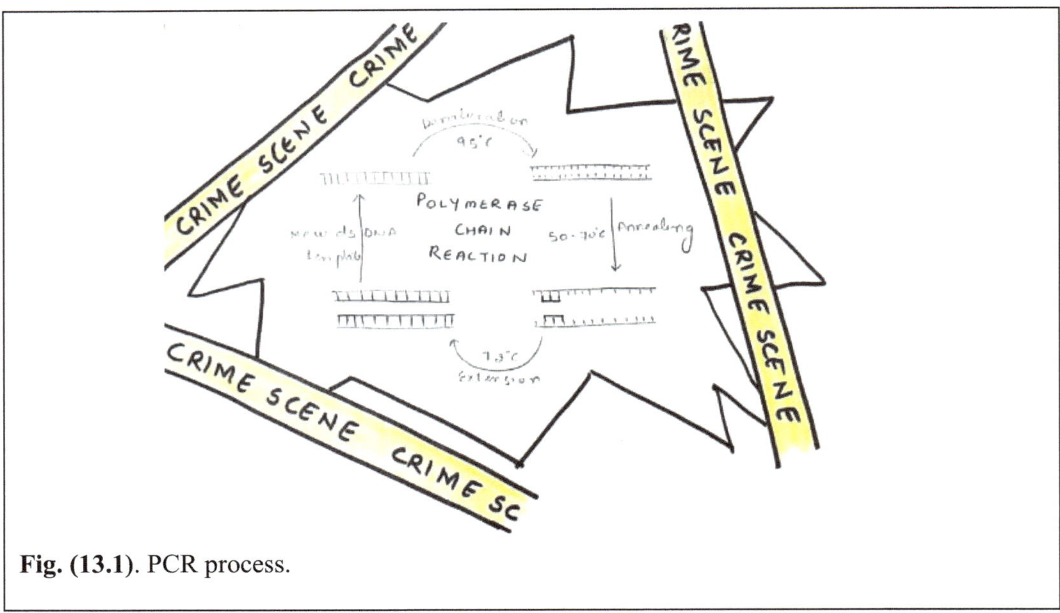

Fig. (13.1). PCR process.

QUESTIONS

1. Write a principle, types and applications of PCR.
2. Describe the vectors for carrying foreign DNA.
3. What is expression vector
4. Write a short note on:
    a. RFLP
    b. Microarray
    c. Palindromic sequences
    d. Genomic library
    e. SNP
    f. VNTR

g. DNA finger printing

h. DNA foot printing

i. Chromosomal walking

j. Restriction endonucleases

k. Tm

5. Difference between genomics and proteomics.

6. Describe the types of stem cells and application of stem cell therapy.

BIBLIOGRAPHY

Denise R Ferrier. Lippincott illustrated reviews: biochemistry. 7th Edition. Philadelphia Wolters Kluwer; 2017

Donald Voet, Judith G Voet, Charlotte W Pratt. Fundamentals of Biochemistry. 5th Edition. New York: Wiley; 2016.

Geoffrey L Zubay, Dennis E Vance. Principles of biochemistry. Dubuque, Iowa: William C. Brown; 1995.

Jeremy M Berg, Gregory J Jr Gatto, Lubert Stryer, John L Tymoczko. Biochemistry. 9th Edition. New York: Macmillan International Higher Education: WH Freeman; 2019.

Keith Wilson, John M Walker. Principles and techniques of biochemistry and molecular biology. 7th edition. Cambridge: Cambridge University Press; 2017.

Lehninger A, Nelson D, Cox M. Lehninger principles of biochemistry. New York: Worth Publishers; 2000.

Michael A Lieberman, Rick E Ricer. Biochemistry, molecular biology, and genetics. 7th Edition. Philadelphia, Pa Wolters Kluwer; 2020.

Victor W Rodwell, David A Bender, Kathleen M Botham, Peter J Kennelly, P Anthony Weil. Harper's illustrated biochemistry. 31st edition. New York: Mcgraw-Hill Education; 2018.

CHAPTER 14

Maintenance of Body Composition

LEARNING OBJECTIVES:	Keywords:
• Describe the sources, biological significance and associated disorders of vitamins and minerals. • Explain the regulation and importance of acid-base, water and electrolyte balance in the body. • Identify and define inherited metabolic disorders. • Categorize xenobiotics and provide their significance. • Illustrate the mechanisms of oxidative stress and antioxidant defences.	Acids, Antioxidants, Bases, Buffers, Deficiency diseases, Dehydration, Electrolytes, Free radicals, Genetic screening, ICF and ECF, IMD, Minerals, pI, pH, ROS, Vitamins,Water, Xenobiotics.

VITAMINS

Sources, Active Form and Importance of Various Vitamins

	Pro Vitamin (Source)	*Active Form*	*Metabolic Importance*
	Fat- soluble Vitamins		
1.	*Vitamin A:*		
	β carotene (vegetable fruit) ↓	Retinal Retinol Retinoic acid	Visual pigment (vision) Coenzyme (sugar transport) Signal molecule (development and differentiation)
	Retinol (Milk, liver, egg yolk)		
2.	Vitamin D	Calcitriol	Hormone

Simmi Kharb

	(Cholesterol → Cholecalciferol (Cod liver oil, milk, egg yolk) ↓ Calcitriol		(Calcium metabolism)
3.	Vitamin E: Tocopherols (Cereals, liver, egg, oil seeds)	Tocopherols	Reducing agent (antioxidant)
4.	Vitamin K: Phylloquinones (Intestinal bacteria, vegetables, liver)	Phylloquinones	Blood clotting
	Water- soluble Vitamins		
	Vitamin B complex:		
5.	Thiamin (grain, yeast)	Thiamin diphosphate (TDP)	Transfer of hydroxyalkyl residues
6.	Riboflavin (milk, egg)	FMN, FAD	Hydrogen transfer
7	Nicotinate, nicotinamide (Meat, yeast, fruit, vegetable)	NAD, NADP	Hydride transfer
8.	Pantothenate (wide distribution)	CoA	Activation of carboxylic acid
9.	Pyridoxal, pyridoxol, pyridoxamine (Meat, vegetable, grains)	PLP (pyridoxal phosphate)	Activation of amino acids

10.	Cobalamin (Meat, liver, egg, milk)	5-deoxy adenosyl cobalamin Methyl cobalamin	Isomerization
11	Biotin (yeast)	Biotin	Transfer of carboxyl groups
	Other Water-soluble Vitamins		
12	Ascorbic acid (Fruit, vegetables)	Ascorbate	Stabilization of enzyme systems, coenzyme, antioxidant.

Fat-Soluble Vitamin: Function and Deficiencies

	Pro Vitamin	*Functional or Active Form*	*Metabolic Importance*	*Features of Deficiency*
1.	Vitamin A Beta carotene Retinol	Retinal Retinol Retinoic acid	Visual pigment (vision) Coenzyme (sugar transport) Signal molecule (development and differentiation)	Defective night vision or night blindness Keratinization of epithelium: Cornea Lungs Conjunctiva GIT Genitourinary tract Bitot's spots Xerophthalmia Blindness Increased susceptibility to infections

2.	Vitamin D (Cholesterol → Cholecalciferol → Calcitriol)	Calcitriol	Hormone (Calcium metabolism)	Rickets (children) Osteomalacia (adults): softening of bone
3.	Vitamin E: Tocopherols	Tocopherols	Reducing agent (antioxidant)	Hemolytic anemia in premature infants, sterility, degenerative changes in muscle and nerves. Neurological symptoms, including loss of muscle coordination, impaired vision and speech.
4.	Vitamin K: Phylloquinones	Phylloquinones	Blood clotting	Bleeding disorders (low prothrombin level, increased clotting time, hemorrhages) Hemorrhagic disease of the newborn

Water-Soluble Vitamin: Function and Deficiencies

Vitamin	*Biochemical Function*	*Features of Deficiency*
Thiamine (B_1)	Cofactor for pyruvate and α ketoglutarate dehydrogenase	Beriberi, heart failure (wet beriberi). Wernicks – korasakoff syndrome: ataxia, ophthalmoplegia, confusion, seen in chronic alcoholics.
Riboflavin (B_2)	Precursor to coenzyme: FMN and FAD	Glossitis (atrophy of tongue,) cheilosis (fissures at corner of mouth), dermatitis, corneal ulceration.
Niacin (B_3)	Required for production of NAD^+ and $NADP^+$ as well as numerous dehydrogenase	Pellagra: diarrhea, dementia, dermatitis. Deficiency can occur as a result of antitubercular drug isoniazid, Hartnup disease, carcinoid syndrome.

Pyridoxine (B_5)	Required for transaminases and decarboxylation reactions	Required for GABA synthesis, so deficiency results in seizures. Deficiency can be associated with isoniazid or pencillamine use.
Biotin	Required for carboxylation reaction	Synthesized by GIT bacteria Deficiency associated with long-term antibiotic use, consumption of raw egg (which contains avidin that binds & inhibits absorption of biotin).
Cobalamin (B_{12})	Required by methyl malonyl CoA mutase and methionine synthase	Deficiency associated with lack of intrinsic factor megaloblastic anaemia, sub acute combined degeneration of spinal cord.
Folate	Reduced by DHF reductase to THF which functions as one carbon donor	Lack of folate results in impaired dTMP synthesis, hence megaloblastic anaemia. Folate supplementation is given in pregnant women for preventing neural tube defects.
Vitamin C	Hydroxylation of proline in collagen Aid in iron absorption	Deficiency results in scurvy.

Vitamin A Deficiency *Symptom*	***Reasoning***
Keratinization	Vitamin A is involved in maintenance of epithelial surfaces. Deficiency leads to epithelial metaplasia in eyes, mucosal membrane, GIT, respiratory tract, genitourinary tract. Mucosa is replaced by inappropriately keratinized stratified squamous epithelium. Also, keratinization of skin occurs termed follicular keratosis.
Night blindness	Reduction in rhodopsin content of rods of retina
Xerophthalmia	Squamous metaplasia of epithelial surface cells become flattened, heaped one upon the other with keratinization of the (surface

<table>
<tr><td></td><td colspan="2">layer) conjunctiva of sclera and cornea of eye. This can lead to softening and destruction of cornea and blindness: Keratomalacia.</td></tr>
<tr><td>Manifestations of Rickets</td><td colspan="2">Reasoning</td></tr>
<tr><td>Wrist, ankle and knee enlarge</td><td colspan="2">Epiphyseal cartilage continues to grow without replacement by bone matrix and mineral</td></tr>
<tr><td>Long bones of leg bow and knee knock (kyphosis)</td><td colspan="2">Occurs as weight bearing activity begin</td></tr>
<tr><td>Spine curved, pelvic and thoracic deformities. Rachitic rosary,</td><td colspan="2">Osteochondral beading at junction of ribs and cartilage</td></tr>
<tr><td>Abdomen distended bossing of frontal and parietal bones</td><td colspan="2">Weak abdominal muscles, delayed closure of anterior fontanelle</td></tr>
<tr><td>Pigeon chest</td><td colspan="2">Undue prominence of sternum bone.</td></tr>
<tr><td>Vitamin E Deficiency May Occur in</td><td colspan="2">Steatorrhoea, abetalipoproteinemia, cholestatic liver disease, cystic fibrosis, intestinal resection, premature infants with inadequate reserve.</td></tr>
<tr><td>Deficiency: Vitamin K</td><td colspan="2">Cause:</td></tr>
<tr><td>*This can be circumvented by giving vitamin K injection deficiency.</td><td>Inadequate intake, absorption of utilization

Fat malabsorption

Pancreatic dysfunction

Biliary disease*</td><td>Atrophy of intestinal mucosa

Steatorrhea of any cause

Sterilization of large intestine by antibiotics, sulfa drugs

Premature infant (inefficient placental transfer)*</td></tr>
</table>

ACID – BASE BALANCE, WATER AND ELECTROLYTE BALANCE

Water and Electrolyte Balance Water is most essential nutrient. Water can move in either direction across cell membrane. Sodium, potassium, calcium, magnesium and phosphate are actively transported, and water follows sodium.	***Fluid*** 50-70% of a healthy adult body is composed of fluids. 2/3rd of body fluid is within cells [intracellular fluid (ICF)]. 1/3rd of body fluid is outside cells [extra cellular fluid (ECF)].	***ECF*** ***Includes*** Tissue (interstitial) fluid found in between cells within tissues and organs of body. Plasma is the fluid portion of blood that carries blood cells.
Functions of Body Fluid	Dissolve and transport substances: carbohydrate, ions, amino acids, vitamins and mineral. Account for blood volume Maintain body temperature	Protect and lubricate body tissues *e.g.* CSF, synovial fluid, amniotic fluid digestive secretions Required for fluid and electrolyte balance and metabolic reaction.
Water and Fluid Balance	***Water Balance*** Body cannot store water. Water lost must be replaced daily. Hypothalamus maintains water balance, regulation of body temperature: thirst is controlled. Sensation of thirst may be impaired in newborns and old age and during disease or extreme dehydration.	***Fluid Balance*** It is maintained by: Replacement of water loss from body Water loss through urine, sweat, exhalation feces Water is gained through beverage, food and metabolic reactions.

<table>
<tr><td rowspan="3">Regulation of Water & Fluid Balance</td><td>Total body water remains constant in health despite massive fluctuations in intake.</td><td rowspan="2">Water excretion by the kidney is tightly controlled by arginine vasopressin (AVP):

Aldosterone and AVP interact to maintain normal volume and concentration of the ECF.</td></tr>
<tr><td rowspan="2">Also, water is lost through the skin in perspiration, and from the lungs during respiration: called 'insensible' loss.

This water loss amounts to between 500 and 850 mL/day.</td></tr>
<tr><td>Water may also be lost in disease: from fistulae, or in diarrhea, or because of prolonged vomiting.</td></tr>
<tr><td rowspan="3">Water and Solute Loss</td><td>Excess body water is eliminated through urine and the extent of urinary salt loss is the main factor that determines urinary water loss.</td><td rowspan="2">A- II (angiotensin II) and aldosterone regulate Na^+ and Cl^- reabsorption in kidney:

They promote Na^+, Cl^- and water reabsorption from kidney in dehydration.</td></tr>
<tr><td rowspan="2">Kidney can respond quickly to meet the body's needs to get rid of water.

Water losses are normally seen as changes in the volume of urine produced.</td></tr>
<tr><td>ANP (atrial natriuretic peptide) promotes natriuresis: excretion of Na+ and Cl- followed by water excretion.</td></tr>
<tr><td>Water and Solute Gain</td><td colspan="2">Body water gain occurs mainly by water intake.</td></tr>
</table>

Causes of Severe Water Loss	***Causes of Dehydration***
Excessive sweating Inadequate consumption of water Repeated vomiting Diarrhea	*CVS*: Hypovolemia: decreased cardiac output, decreased BP, vasoconstriction, Hemolysis, DIC (disseminated intravascular coagulation) *Kidney*: Decreased GFR, decreased renal blood flow, decreased urine flow decreased urine specific gravity.

Electrolytes: Na, K, Cl	Blood fluid is composed of: Water Electrolyte: Na, K, Cl, P (HPO_4^{2-}, PO_4)	Sodium and potassium are cations with sodium in greater concentration extracellularly and potassium in greater concentration intracellularly. Chloride and bicarbonates are anions.	
Function of Electrolytes	Regulate fluid balance: Water follows movement of electrolyte, moving by osmosis.	Help nerves respond to stimuli: Electrolytes (Na, Cl) generate action potential across membrane.	Help in muscle contraction: *Via* movement of calcium.
Sodium	Total body sodium of average 70 kg man: Approximately 3700 mmol Approximately 75% of which is exchangeable.	Sodium is dominant cation of ECF. NaCl provides 90-95% of ECF osmolarity. Sodium levels in ECF remains stable, sodium concentration is tightly regulated at around 140 mmol/L.	
Sodium Levels in Body are Maintained by	Renal reabsorption in proximal convoluted tubule.	Aldosterone *Via* DCT in response to blood volume and blood pressure.	
Sodium Loss	Sodium may be lost from the body in urine or from the gut	*E.g.* prolonged vomiting, diarrhea and intestinal fistulae.	
Abnormal Sodium Concentration in ECF	***Hypernatremia*** It is caused by renal severe diarrhea, burns or excessive sweating, nephrogenic diabetes insipidus.	***Hyponatremia*** Clinically significant hyponatremia is relatively uncommon and is nonspecific in its presentation	

Potassium	It is major intracellular cation.	It is controlled by Na-K-ATPase pump.
Function	It is dominant cation in ICF.	Potassium maintains cardiac rhythm and contributes to neuromuscular conduction.
Potassium Balances in ICF	98% of K is in ICF. Cells spend energy to recover K ions diffused from cytoplasm into ECF. ECF volume is monitored by baroreceptors at carotid sinus, aortic sinus, and right atrium. Serum potassium concentration is normally kept within a tight range (3.5–5.3 mmol/L).	
Abnormal Potassium Levels	***<u>Hyperkalemia</u>*** It may be caused by decreased renal excretion in acute or chronic renal failure, certain diuretics, hypoaldosteronism. It may occur with ion shifts caused in diabetic ketoacidosis, leukemia, excessive muscle activity, and hemolysis.	***<u>Hypokalemia</u>*** It is caused by renal tubular acidosis, hyperaldosteronism, diuretics, vomiting, diarrhea, laxatives.
Chloride (Cl)	It is prevalent anion in ECF. It moves easily across ICF and ECF *Via* channels and transporters. Cl is regulated by ADH which governs water loss. Processes that cause increase or decrease in renal reabsorption of Na^+ affect the Cl reabsorption. Chloride maintains acid base balance by maintaining H ion concentration	
Acid – Base Balance: Terminology		
Proton	H^+ or H_3O^+ : a hydrogen nucleus without electrons	In water, RNH_3^+ is acid and RNH_2 is not. Strong acids completely dissociate in water, weak acids do not.

<table>
<tr><td>id</td><td>Has a proton, gives up a proton</td><td rowspan="2"></td></tr>
<tr><td>Base</td><td>Takes up a proton</td></tr>
</table>

<table>
<tr><td>Strong acid at equilibrium dissociate into:
$HCl + H_2O \rightleftharpoons H_3O^+ + Cl^-$
Strength of an acid is defined as its tendency to give up its proton to water. Stronger the acid is, larger is the equilibrium constant (Ka) for the reaction.
$HA + H_2O \rightleftharpoons A^+ + H_3O^+$
$Ka = \frac{[A^-][H_3{}^+O^-]}{[HA]}$
For strong acids, Ka > 1</td><td>For weak acids, Ka < 1
pKa = -log (Ka)
pH = -log $[H^+]$
The lower the pH, the higher the $[H^+]$
pKa = $-\log_{10}$ (Ka)
➢ Lower the pKa (higher Ka)→ stronger the acid</td></tr>
</table>

<table>
<tr><td>Buffers</td><td>Buffers are solutions that contain both acidic and basic forms of a weak acid. Buffers minimize changes, in pH when strong acids and bases are added.</td></tr>
</table>

<table>
<tr><td>Henderson – Hassel batch equation:
$pH = pKa + \log \frac{[base]}{[acid]}$
[base] > [acid] ⇒ pH > pKa
The key to understanding buffers is in the ratio of [base] / [acid].
If ratio = 1, log [base] / [acid] = zero and pH = pKa.</td><td>PI = isoelectric point
pH = PI ⇒ no net charge
pH > PI ⇒ negative charge
pH < PI ⇒ positive charge
About half way between titration of carboxyl and amino group, the molecule must have no net charge. At PI, molecule is neutral.</td></tr>
</table>

Types of Buffers	Bicarbonate buffer	Phosphate buffer	Protein buffer

Bicarbonate Buffer

CO_2 - bicarbonate buffer is important buffer in blood. Acid form of bicarbonate buffer is actually a gas dissolved in water. Dissolved CO_2 gets hydrated to form acid H_2CO_3 which gives up a proton to a base and makes HCO_3^-.

$$CO_2 + H_2O \rightleftharpoons H_2CO_3 \rightleftharpoons H^+ + HCO_3^-$$

So, CO_2 is acid and HCO_3^- is base in blood and the effective pKa is 6.1.

If HCO_3^- (base) > CO_2 (acid), then pH > pKa

At physiological pH 7.4, pKa 6.1, HCO_3 is tenfold more than CO_2. So, now bicarbonate buffers effectively at 7.4 when its pKa is 6.1.

CO_2 is exhaled from lungs and proton is not exhaled and first left behind and turned into water.

$$H^+ + HCO_3^- \rightleftharpoons CO_2 + H_2O$$

CO_2 concentration is expressed in pressure units of mmHg and denoted at pCO_2. pCO_2 is 40mmHg and atmospheric pressure is 760 mmHg which means that CO_2 represents 40/760 of total pressure of gas above liquid *i.e.*, ~ 5%.

Actual concentration of dissolved CO_2 (dCO_2 in millimoles / l) is =

dCO_2 (in millimoles / l) = 0.03 pCO_2 (in mmHg)

$$Ka = \frac{[H^+][dCO_2]}{[HCO_3^-]}$$

$Ka = 10^{-6.1} = 7.94 \times 10^{-7}$ M

$dCO_2 = 0.03\ pCO_2$

$$\text{So, } [H^+] = \frac{24pCO_2}{[HCO_3^-]}$$

How CO_2 Concentration is Regulated	CO_2 concentration is regulated by lungs and $[HCO_3^-]$ is regulated by kidneys.
Lungs	Lungs can control the CO_2 concentration in blood by changing the respiratory rate. *Hyperventilation (rapid, deep respiration)*: Causes more CO_2 to leave the body causing fall in pCO_2 and thus pH increases. *Hypoventilation (slow, shallow respiration):* Causes blood pCO_2 to rise which decreases pH.
Kidneys	Kidneys regulate $[HCO_3^-]$ concentration: Kidneys decide to reclaim the carbonate (on the basis of pH of initial filtrate) back in blood or let its loss from body. If bicarbonate is not reclaimed, then serum bicarbonate concentration falls If bicarbonate is reclaimed more than normal, bicarbonate concentration rises. ➢ *So, to keep pH very close to 7.4, CO_2 and HCO_3 are removed at exactly the same rate as they are formed.* Kidneys reclaim HCO_3 to maintain pH Kidney secrete $[H^+]$ directly into filtrate, lowering pH of urine. So, filtered bicarbonate exists as CO_2. CO_2 is a freely diffusible gas across membranes and loss of H^+ results in formation of bicarbonate and reclamation of HCO_3 and proton. This excess proton exits the body on another buffer in urine (usually phosphate buffer). *Net result* is fall in urine pH and kidney reclaims more of bicarbonate which causes rise in $[HCO_3]$ and restoration of pH to normal. $[H^+] = \frac{24pCO_2 \uparrow}{[HCO_3^-] \uparrow}$

	Thus, increased pCO_2 caused by a problem with lungs is compensated by kidneys by reclaiming HCO_3.

Normal pH Balance and its Derangement

Normal pH balance

Lungs have numerator (pCO_2)

Kidneys have denominator ($[HCO_3^-)]$

The amount of CO_2 produced is balanced by its loss from lungs and kidneys.

If pH is changed; it implies that the capacity of the system must have been exceeded.

1. Either by abnormally high pCO_2 (increased numerator)
2. Or by abnormally low $[HCO_3^-]$ (decreased denominator)

In this equation:

$[H^+] = \frac{24pCO_2}{[HCO_3^-]}$

Two things can cause abnormal pH =

Change in pCO_2 or

Change in $[HCO_3^-]$

And 24 never changes!

Acidosis: If $[H^+]$ is high (low pH)

Alkalosis: If $[H^+]$ is low (high pH)

For each type of $[H^+]$ imbalance, there are two possible causes:

As *pCO_2* is regulated by lungs, label respiratory is attached to effects that cause pCO_2 to change *i.e.*, respiratory acidosis or alkalosis (pCO_2 ↑ or ↓).

	pH	Type
Acidosis	<7.4	Respiratory pCO_2 ↑ Metabolic $[HCO_3^-]$↓
Alkalosis	>7.4	Respiratory pCO_2 ↓ Metabolic $[HCO_3^-]$↑

Respiratory Alkalosis	***Respiratory Acidosis***
pCO_2 decreases, $[H^+]$ decreases	pCO_2 increases, $[H^+]$ increases.

As $[HCO_3^-]$ is regulated by kidney so changes in bicarbonate, concentrations are labeled 'metabolic'.

Metabolic Acidosis	***Metabolic Alkalosis***
If HCO_3 drops	If bicarbonate rises

Compensatory Mechanisms	When the acid-base balance breaks down, the body has two options: either to fix the original problem or provide a temporary fix till the permanent fix comes into action(Fig. **14.1**).
Hypoventilation raises pCO_2, decreases pH ($[H^+]$ increased) causing respiratory acidosis.	***Compensation:*** hyperventilation does not seem to be an option. Strategy should be to change bicarbonate concentration to get pH back to normal and this becomes metabolic alkalosis.

Condition	**PCO_2**	**H+**	**$[HCO_3^-]$**	**Compensation**
R. Acidosis	↑	↑	↑	M alkalosis
R. alkalosis	↓	↓	↓	M acidosis
M. acidosis	↓	↑	↓	R. alkalosis
M. alkalosis	↑	↓	↑	R. acidosis

Thus, metabolic cause is compensated by respiratory, acidosis is compensated by alkalosis!

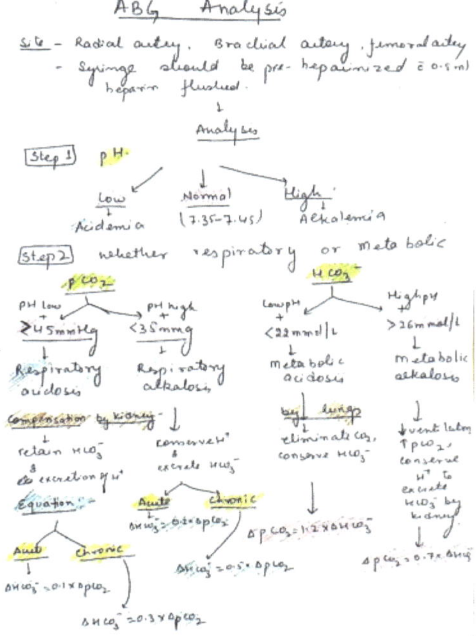

Fig. (14.1). ABG Analysis showing derangements in pH balance.

Clinical Causes of Acid- Base Disorders			
Metabolic		***Respiratory***	
Acidosis	***Alkalosis***	***Acidosis***	***Alkalosis***
Diabetes mellitus(ketoacidosis)	Vomiting (loss of H^+)	COPD (chronic obstructive airway disease)	Hyperventilation (anxiety, fever)
Lactacidosis	Nasogastric suction (loss of H^+)	Cardiac arrest	Salicylate poisoning
Renal failure		Respiratory	
Diarrhea (Loss of HCO_3)	Hypokalemia	Central depression (drugs: opiate)	Anemia
Renal tubular acidosis (HCO_3 loss, H^+ impairment)	IV HCO_3 administration	Airway obstruction	

MINERAL METABOLISM

Calcium	***Extracellular Calcium is Present as***
Calcium is fifth most abundant element. Human body contains 1-1.5 kg calcium, most of which is located in bones (98%) and rest is extracellular.	Free (ionized 47%) form Protein bound (47%: 80% with albumin and 20% with globulin) Complexed with organic acids (6% bound to citrate phosphate, bicarbonate and lactate).

<table>
<tr><td colspan="2">Functions

Calcium is structural component of bone and teeth.

Signaling: It functions as a signaling molecule, triggers exocytosis, muscle contraction.

Acts as cofactor in coagulation of blood.</td><td colspan="2">Calcium Homeostasis

Calcium metabolism is balanced in healthy adults.

Approximately 1g Ca^{+2} is taken up daily of which 300 mg is absorbed.

Maintenance of calcium balance depends on efficiency of intestinal absorption, regulation by parathyroid hormone, (PTH) vitamin D and calcitonin.</td></tr>
<tr><td colspan="4">If ionized calcium levels fall, nerves become hyper excitable and may develop tetany.</td></tr>
<tr><td colspan="2">Regulation

Calcitonin counter acts the action of PTH and lowers calcium levels.

Calcium is absorbed in intestine and acidic pH; high protein diet favors its absorption.</td><td colspan="2">Oxalates, phytates, fiber, excess phosphate, magnesium, iron inhibit its absorption.

About 100-400 mg calcium is excreted per day.

Serum calcium levels are maintained between 8.8-10mg/dL</td></tr>
<tr><td>Disorders of Calcium</td><td colspan="2"><u>Hypercalcemia</u></td><td><u>Hypocalcemia</u></td></tr>
<tr><td><u>Cause</u></td><td colspan="2">PTH: primary hypoparathyroidism

Malignancy: breast, liver, kidney

Vitamin D intoxication

High bone turnover (hyperthyroidism)

Renal failure</td><td>PTH absent or in effective

Chronic renal failure lack of vitamin D</td></tr>
</table>

Features	Bone pains, fracture Lethargy, confusion, coma constipation, nausea, vomiting Kidney stones Short QT interval, broad T waves	Neuromuscular irritability Numbness Tingling cramps tetany.
Rickets	It is disorder of defective mineralization of skeleton due to vitamin D deficiency commonly. It presents as skeletal deformity (craniotabes, rachitic rosary), increased susceptibility to fractures, bowing of legs, delayed eruption of teeth.	
Phosphorus Phosphorus is most abundant intracellular anion. Serum phosphorus levels are 2.5-4.5 mg/dl.	Total body phosphorus is 700g of which: 85% (600g) is in bones 15% in soft tissue 0.1% in extracellular fluid.	***Phosphorus Absorption*** About 1g phosphorus is taken in diet daily and gets absorbed in intestine. Phosphorus absorption is under influence of vitamin D and excretion is under PTH.
Function	Phosphorus is important constituent of bone and teeth. It is important component of coenzyme (NADP, TPP, PLP), ATP and blood phosphate buffer.	
Disorders of Phosphorus	***Hyperphosphatemia***	***Hypophosphatemia***
Cause	Hemolysis Increased release by cells Increased release from cells Malignancy renal failure	Decreased Intake Increased excretion (divretics) Increased uptake by cells.

<table>
<tr><td>Features</td><td>Ectopic calcification</td><td>Anorexia

Dizziness

Bone pain

Muscle weakness

Waddling gait</td></tr>
<tr><td>Iron</td><td>Peculiarities

Fourth most abundant element in earth crust

Iron is one of the most essential trace elements in the body</td><td>Total body iron content is 3.5 g:

70% is in blood in RBCs as constituent of Hb

5% in myoglobin of muscle (rest is in liver, bone marrow, muscle)</td></tr>
<tr><td colspan="3">Iron is unique in that it operates largely as a closed system with iron stores being efficiently reutilized by the body</td></tr>
</table>

<table>
<tr><td>Biochemical Functions of Iron

1. Exerts its function through compounds in which it is present:

Hb and Myoglobin required for transport of O_2 and CO_2.

2. Cytochromes and other non-heme iron necessary for ETC and oxidative phosphorylation.

3. Peroxidase required in phagocytosis of bacteria in neutrophil.</td><td>Unique: Homeostasis is maintained by regulation at level of absorption and not excretion.

Iron is a One-Way

• It is unique, efficiently substance-utilized and reutilized in the body.

• Iron losses from body are minimum.

• Not excreted in urine.

• Not inactivated (like vitamin and other substance) during the course of metabolic function.</td></tr>
<tr><td colspan="2">When stores are adequate – absorption decreased. When stores are depleted – absorption increased. This is referred to as MUCOSAL BLOCK theory.</td></tr>
</table>

<table>
<tr><td colspan="3">Iron from intestinal lumen enters mucosal cells in ferrous state</td></tr>
<tr><td colspan="2">Disorder of Iron Metabolism

<u>Iron -Deficiency</u>

Inadequate Intake

Impaired absorption, MAS, surgery, excessive loss, menstrual loss, GI loss, post gastrectomy anemia.

Ineffective erythropoiesis, kidney disease, hook worm infestation, pregnancy</td><td>Lab findings:

Microcytic, Hypochromic anaemia

Hb< 12 mg/dl

Mechanism:

Depletion of Fe stores cause Fe deficient erythropoiesis resulting in Fe deficient anaemia.</td></tr>
<tr><td colspan="2" rowspan="2">Causes of Fe- Deficiency Anaemia:

Hookworm infection

Nutritional deficiency of iron

Repeated pregnancy

Chronic blood loss: piles, bleeding, peptic ulcer, uterine hemorrhage.

Kidney disease - nephrosis, CRF

Lack of absorption – gastrectomy (partial), achlorhydria

Lead toxicity</td><td>Signs and Symptoms

Weakness, lassitude, apathy, sluggish metabolic activities, retarded growth, loss of appetite, palpitation exertional dyspnoea, koilonychia, pica.</td></tr>
<tr><td>Treatment

Iron and folic acid supplementation.</td></tr>
<tr><td>Iron Toxicity</td><td><u>Hemosiderosis</u></td><td><u>Haemochromatosis</u></td></tr>
<tr><td></td><td>- Iron excess

- Seen in subject receiving repeated blood transfusion.</td><td>Primary:

Inherited autosomal recessive</td></tr>
</table>

	E.g. Hemophilcs, thalassemia, Bantu siderosis seen in African Bantu Tribe. (due to habit of cooking iron in iron pots).	Abnormal gene on short arm of chromosome 6 Here iron absorption is increased and transferrin levels are increased and excess iron is deposited
Treatment	phlebotomy desferroxamine – chelating agent: chelates Fe^{+3} and excreted in urine	Fe gets deposited in liver – cirrhosis Pancreas – DM Skin – discoloration

Magnesium

Fact File	***Magnesium Functions***
Most abundant intracellular cation Total body Mg 0.3 g/kg body weight 1% extracellular 31% in cells 67% in bones Serum magnesium: 2-3 mg/dl 1-4mg/dl Mg is in unbound, diffusible form	1. Enzyme action – as cofactor (activator) of many enzymes requiring ATP *E.g.*, H.K., GK, FK, PFK, Adenyl cyclase, ALP: cAMP dependent kinases. Most Mg inside cell is bound to ATP: Mg- ATP is in equilibrium with free Mg ions. 2. Neuromuscular excitability (lower muscle irritability): action similar to that of Ca^{+2} 3. It is required for formation of bone and teeth. 70% present as apatite in bone, dental enamel, and dentin. 4. Magnesium is an important cofactor for carboxylase and is involved in DNA replication and RNA synthesis.

<table>
<tr><td>Ideal intake 36-48 mg/day</td><td colspan="2"></td></tr>
<tr><td>Plasma magnesium concentration is 0.6-1-0 mmol/l and intracellular concentration is 10mmol/l.

When Mg intake is restricted, fecal excretion becomes negligible and urinary excretion becomes negligible. About 70-80% of Mg is filtered by kidney and 3-5% excreted.</td><td colspan="2">Regulation: Occurs at following levels:

(i) Mainly by kidney: Mg retention by kidney is very efficient.

(ii) Absorption from GIT

(iii) Transmembrane fluxes</td></tr>
<tr><td colspan="3">Magnesium Deficiency manifests as muscle weakness, failure to thrive, neuromuscular dysfunction, ventricular tachycardia, coma and death.</td></tr>
<tr><td>Cause</td><td>Inadequate intake

Chronic alcoholism</td><td>GIT disorders:

Malabsorption syndromes

Prolonged diarrhea</td></tr>
<tr><td>Mg deficiency impairs normocalcemic response to PTH at skeletal level.</td><td>Endocrine:

Hyper/hypoparathyroidism

Hyperthyroidism

Diabetic ketoacidosis

SIADH</td><td>Increased renal excretion:

Alcohol ingestion

Cisplatin therapy

Diuretics: furosemide</td></tr>
<tr><td>Increased Mg Levels are Found in</td><td colspan="2">Renal failure, uncontrolled DM, eclampsia, nausea, sedation, decrease deep tendon reflexes, muscle, weakness, hypotension, coma, respiratory paralysis.</td></tr>
<tr><td>Copper

Serum Copper</td><td colspan="2">100 μg/dl in plasma

90-95% tightly bound to ceruloplasmin</td></tr>
</table>

<table>
<tr><td>Total body copper is 50-80mg in an adult and is mainly concentrated in muscle and liver.</td><td colspan="2">5-10% bound to albumin

1% bound to amino acids (histidine), transport form of copper</td></tr>
<tr><td colspan="3">Normally ceruloplasmin in serum is 25-50mg/dl, also known as ferroxidase: promotes oxidation of Fe^{+2} to Fe^{+3}.</td></tr>
<tr><td colspan="3">Function: Copper is constituent of a wide variety of enzymes:</td></tr>
<tr><td>1. Ceruloplasmin:

(i)Mobilization of iron (also known as ferroxidase I) and utilization

(ii) Acute phase protein: increased in infection and inflammation

(iii) It is not a copper transport protein, but it binds copper tightly

(iv) It acts as a free radical scavenger.</td><td>2. Cytochrome C oxidase

3. SOD (superoxide) dismutase): dismutation of O_2 (superoxide) to H_2O_2

4. Lysyl oxidase: oxidizes four lysine residues to desmosine which makes cross linkages in collagen and elastin.

5. Dopamine β-hydroxylase: converts dopamine to norepinephrine.

6. Tyrosinase: converts tyrosine to DOPA.

7. Catalase</td><td>8. Others:

Tryptophan pyrrolase

Uricase

δ-aminolaevulinic acid (δALA): Heme synthesis

MAO (mono amine oxidase)

Ascorbic acid oxidase

9. Phospholipid synthesis, development of bone and nervous system.</td></tr>
<tr><td colspan="3">Disorder of Copper</td></tr>
<tr><td><u>Copper Deficiency</u>

<u>Cause:</u> MAS, nephrotic syndrome, celiac disease.</td><td><u>Clinical Features</u>

Weight loss, demineralization of bones, defect in collagen formation, and demyelination of nerves.</td><td><u>Lab Findings</u>

Hypochromic microcytic anaemia, fragility of blood vessels, aneurysms, hypopigmentation, myocardial fibrosis and grey hair.</td></tr>
</table>

<table>
<tr>
<td><u>2. Inherited Disorders</u>

(a) Wilson's disease
(Hepatolenticular degeneration)</td>
<td>Autosomal recessive disorder in which copper is deposited in abnormal amount in liver, lenticular nucleus of brain. Serum copper and serum ceruloplasmin (cp) levels are low and increased excretion of copper occurs in urine. Copper deposition in kidney</td>
<td>Cause:

There is defect in incorporation of copper into newly synthesized apoceruloplasmin. Hepatic lysosomes lack normal mechanism to excrete the copper that has been cleaved from ceruloplasmin (cp) into bile, and copper excess inhibits formation of Cp from apo Cp and copper. When capacity of hepatocytes exceeds to store copper, then it is released into circulation and its uptake in extrahepatic sites occurs.</td>
</tr>
<tr>
<td>Result of Copper Excess</td>
<td>Copper gets deposited in various tissue.

Copper deposition in liver causes hepatic cirrhosis.

CNS effects: necrosis of neurons and sclerosis

Kidney: renal tubular absorption defects</td>
<td rowspan="2"><u>Diagnosis</u>

Cp < 20 mg%

KF ring

Hepatic cirrhosis

Urine copper excretion: markedly increased.</td>
</tr>
<tr>
<td colspan="2">Eyes: Kayser Fleischer ring (KF ring) – copper deposition occurs in Descemet's membrane and golden green ring is seen (on slit lamp examination).</td>
</tr>
<tr>
<td>(b) Menke's Disease

Kinky or Steel Hair Syndrome

X-linked, affects males only efflux of copper through basolateral membrane is defective.</td>
<td colspan="2"><u>Clinical Features</u>

Signs of mental retardation abnormal bone formation, anaemia, depigmentation of hair, no deposition of copper, low plasma ceruloplasmin.</td>
</tr>
</table>

<table>
<tr><td colspan="2"></td><td><u>Treatment</u>

Penicillamine – chelating agent.</td></tr>
<tr><td colspan="3">Zinc</td></tr>
<tr><td rowspan="3">Function</td><td rowspan="3">1. Role in enzyme function:

Metalloenzymes (up to 300):

i.SOD (super oxide dismutase)

ii.Carbonic anhydrase (present in RBC and kidney).

iii.DNA-dependent DNA polymerase

iv.DNA dependent RNA polymerase

v.ALP (alkaline phosphatase)

vi.Carboxypeptidase (pancreatic)

vii. Leucine aminopeptidase

viii LDH

ix.Alcohol dehydrogenase

x. ALA synthetase.</td><td>2. Role in vitamin A metabolism:

Zinc stimulates vitamin A release from liver and increases its plasma level and its utilization in rhodopsin synthesis.

Retinine reductase is zinc containing metalloenzyme which participates in regeneration of rhodopsin in eye during dark adaptation.</td></tr>
<tr><td>3. Role in Insulin secretion:

Important for synthesis and storage: insulin in beta cell contains zinc and zinc helps in release of insulin.</td></tr>
<tr><td>4. In growth and development

5. In wound healing

6. Role in biosynthesis of nucleotides</td></tr>
<tr><td>Disorder of Zinc</td><td><u>Deficiency</u>

Zinc deficiency is rare since it is widely available in foods.</td><td>Causes of Low Serum Zinc Levels

-Decreased intake or absorption – regional enteritis, malabsorption</td></tr>
</table>

<table>
<tr><td rowspan="2"></td><td>Iatrogenic: use of antimetabolites and diuretics. Acrodermatitis entero pathica (associated with dermatitis, alopecia, diarrhoea)</td><td colspan="2" rowspan="2">-Increased urinary loss: nephrotic, cirrhosis, hypoalbumin states

-Hypercatabolic states: trauma, burns, surgery, hemolytic anaemia, sickle disease

-Decreased with infection, hepatitis, malignancy

-Severe zinc deficiency is seen in: alcoholics (especially if they have liver cirrhosis)</td></tr>
<tr><td>Clinical features:

In children, zinc deficiency causes growth retardation, infections and diarrhoea. In adults, it produces loss of appetite, diarrhoea, anaemia, skin erosion and alopacia.</td></tr>
<tr><td></td><td>Growth retardation

Poor wound healing

Anaemia

Alopecia

Tissues with a high cellular turnover are characteristically affected.

Hyperkeratosis, dermatitis over knee, elbow.</td><td colspan="2">Skin, GIT mucosa, chondrocytes, spermatogonia, thymocytes

Diarrhea

Immunological dysfunction/defects of T- cell function: superinfections common

Gonadal atrophy

Impaired spermatogenesis

Congenital malformations</td></tr>
<tr><td rowspan="2">Toxicity (Excess)</td><td rowspan="2"><u>Sources of Excess</u>

- Inhalation of zinc fumes (welders): metal fumes, fever or brass chills

- Oral ingestion, IV administration</td><td colspan="2">Features

It can lead to nausea, vomiting and fever.</td></tr>
<tr><td>Clinical features:

Gastric ulcer

Pancreatitis

Lethargy</td><td>Fever

Nausea, vomiting

Respiratory distress

Pulmonary fibrosis</td></tr>
</table>

<table>
<tr><td></td><td>- Contamination of dialysis fluid with Zn from adhesive plaster</td><td>Anaemia</td><td></td></tr>
<tr><td></td><td>Clinical features:
Acrodermatitis (lips and tips)
Diarrhoea</td><td colspan="2">Alopecia
GIT
Ophthalmic, neuropsychiatric features.</td></tr>
</table>

<table>
<tr><td>Manganese (Mn)</td><td colspan="3"></td></tr>
<tr><td>Function
1. Role in enzyme action: Manganese acts as an activator of enzymes and component of metalloenzyme.
<u>Co-factor:</u></td><td>Arginase
Glutamine synthase
Isocitrate dehydrogenase
Succinate dehydrogenase
Enolase</td><td>Hexokinase
Phosphoglucomutase
Lipoprotein lipase
Metalloenzyme:</td><td>Mitochondrial SOD
Pyruvate carboxylase
Acetyl CoA carboxylase
Diamine oxidase</td></tr>
</table>

<table>
<tr><td>2. Important for synthesis of glycoproteins and chondroitin sulphate (as it is integral part of glycosyl transferase) required for bone formation and function of nervous system.
3. Porphyrin synthesis: (δ ALA synthetase)</td><td>4. Condensation of mevalonate of squalene is Mn dependent: in cholesterol biosynthesis.
5. In antioxidant property of vitamin E
6. Required for skeletal growth</td></tr>
</table>

<table>
<tr><td>Disorder of Manganese</td><td><u>Deficiency</u>
Rare
Retarded growth and skeletal deformities (increase ALP)</td><td><u>Toxicity</u>
Miners inhaling manganese dust develop: anorexia, apathy, headache, leg cramp, psychosis and pneumoconiosis.</td></tr>
</table>

	Bleeding disorder (increase PT time) Fatty liver – Decrease activity of cells of pancreas as in manganese deficiency pancreatic hyperplasia is reported.	
Aluminum (Al)	Aluminum is found in all biological materials and high concentrations are found orange juice. Also found in herbs, processed cheese, baking powder and pickles. Daily intake is 20mg/kg of food. Fluoride and silicon reduces bioavailability of iron and it is excreted in urine.	
Biological Role Only known biological role is it is part of succinate dehydrogenase-cytochrome C system	***Disorder of Aluminium*** Implicated in Alzheimer's disease, dementia following renal dialysis if dialysis fluid contains aluminum.	
Antimony	It does not appear to be essential element and daily intake is 2-10 μmol/day in diet. It is absorbed from diet and efficiency is 15% and it is stored in liver, kidney, skin and adrenals.	
Disorder of Antimony Toxic effects include GIT and respiratory symptoms		
Fluoride		
Function 1. ***<u>Role to Tooth Development and Dental Health</u>*** Fluoride is present in human tooth in trace amounts and helps in tooth development, normal maintenance and hardening of dental enamel. It has a role in prevention of dental caries. Cariostatic effect of fluoride is due to its entry into apatite and salts of dental enamel: <0.5 ppm fluoride in water predisposes to caries.	***<u>2. Role in Bone Development</u>*** ***Disorder of Fluoride*** ***<u>High Fluoride Intake</u>*** High intake of fluoride in excess of 1mg/l result in mottling of teeth. Increased fluoride in blood and tissues cause increase content of bone stimulates osteoblasts activity. There is abnormal rise in calcium deposition: osteosclerosis, calcification of ligaments and tendons. Crippling deformities such as kyphosis, stiffness of spine, bony exostosis occur	

> 1.2 ppm fluoride in drinking water increase dentine:	
Cobalt (Co) ***Function*** Cobalt is component of vitamin B_{12} and this is the only biological function of the element: 1.Role in cobamide enzyme formation *i.e.*, adenosyl cobalamin. 2.Required in methionine metabolism: enzyme transferase *e.g.* homocysteine methyltransferase 3.Bone marrow: development, maturation of RBCs.	***Disorder of Cobalt*** ***Deficiency:*** Anaemia (B_{12} deficiency) ***Toxicity*** Prolonged administration causes polycythemia, cardiomyopathy, CHF (congestive heart failure) with pericardial effusion polycythemia. Thyroid enlargement and neurological abnormalities also have been reported as manifestation of cobalt toxicity in drinkers of beers as metal is added as foam.
Selenium ***Se Levels*** 0.2 μg/g in blood In tissue: 13 μg% Se intake depends on nature of the soil in which crops are grown.	***Function*** 1. Function in metalloenzyme: GSH peroxidase (GSH Px) which destroys peroxides in cytosol. *i.e.*, plays a critical role in control of O_2 metabolism, particularly in breakdown of H_2O_2.

<table>
<tr><td colspan="2">2. Present as:

Selenomethionine

Selenocysteine

Selenocystine</td><td colspan="2">3.Antioxidant and protects cells against FR (free radicals) – prevent lipid peroxidation

4.Sparing effect on vitamin E and reduces vitamin E requirement

5.Selenocysteine is component of GSH Px</td></tr>
<tr><td colspan="2">6.Binds heavy metals: Cd, Hg and Ag and protects from their toxic effects 5' deiodinase is Se containing enzyme.</td><td colspan="2">7. Necessary for normal development of spermatozoa

8. Protects animals from carcinogenic chemicals, a role in man is not established.</td></tr>
<tr><td>Disorder of Selenium</td><td colspan="2"><u>Deficiency</u>

There is overlap between selenium deficiency and vitamin E deficiency. Selenium deficiency is associated with cardiomyopathy, CHF, muscle degeneration</td><td><u>Toxicity</u>

Alopecia

Abnormal nails, emotional lability, lassitude.</td></tr>
<tr><td><u>Keshan's Disease</u>

Multifocal myocardial necrosis and decreased serum selenium content due to deficiency of Se in soil and diet in Keshan province (China).</td><td colspan="3">Clinical Features

Arrhythmia, carcinogenic shock, peripheral myopathy.

Young women and children are particularly susceptible.</td></tr>
<tr><td>Acute Poisoning</td><td colspan="3">Infestation of water containing large amount of metal result in diarrhoea, tetanic spasm, respiratory failure.</td></tr>
<tr><td>Chronic Poisoning</td><td colspan="3">Anorexia, hair loss, myocardial atrophy and liver necrosis in animals.

Selenium is present in metal polishes and antirust compound. In human, chronic dermatitis, alopecia, brittle nails, garlicky breath</td></tr>
</table>

<table>
<tr><td></td><td colspan="2">(because of dimethyl selenide) is seen in occupational exposure to electronics, glass and paint industries.</td></tr>
<tr><td colspan="3">Others</td></tr>
<tr><td colspan="2">Iodine

<u>Function</u>

Iodine is required for synthesis of thyroid hormone: T_3, T_4

<u>Dietary requirement:</u>

100-150 μg/d, in pregnancy + 50 μg/d</td><td><u>Disorder of Iodine</u>

<u>Deficiency</u>

Iodine deficiency results in thyroid enlargement, namely goitre. Iodine deficiency has consequences for normal growth and development.

Iodine deficiency causes reduced thyroid iodine stores, reduced production of thyroid hormones and increased production of TSH (release of feedback inhibition).</td></tr>
<tr><td colspan="2"><u>Cause</u>

Iodine deficiency can be due to dietary deficiency of iodine, consumption of plant goitrogens (thiocyanate).</td><td><u>Manifestation</u>

Iodine deficiency in children causes goitre, myxedema (refer to endocrines chapter for further details).</td></tr>
<tr><td colspan="2">Arsenic

Arsenic is present in many plants and animal foods and yet to be established as an essential element in diet. Arsenic is absorbed and excreted in urine and bile</td><td>Disorder of Arsenic

Arsenic contamination occurs Via water, poultry, pig feed muscle and liver and sea food.</td></tr>
<tr><td>Boron</td><td colspan="2">It forms complexes with sugars, polysaccharides, adenosine phosphate, pyridoxine, riboflavin, vitamin C and steroid hormones. It is an essential nutrient. It is found is plants and excreted in urine. There is no evidence of its role in metabolism.</td></tr>
<tr><td>Bromine</td><td colspan="2">It does not have any physiological function. It can concentrate in thyroid tissue. Its deficiency is unlikely to occur.</td></tr>
<tr><td>Cadmium</td><td colspan="2">It is not an essential element and exposure occurs from industrial waste contaminated water, cigarette smoke.</td></tr>
</table>

<table>
<tr><td></td><td colspan="2">Disorder of Cadmium

Cadmium exposure occurs from industrial waste contaminated water, cigarette smoke.</td></tr>
<tr><td>Caesium</td><td colspan="2">It is available as radioisotope caesium 137. It is metabolism in a manner similar to potassium.</td></tr>
<tr><td>Molybdenum (Mo)</td><td><u>Function</u>

<u>Enzyme action:</u>

Xanthine oxidase

Helps in utilization of copper

Aldehyde oxidase.</td><td><u>Disorder of Molybdenum</u>

<u>Molybdenosis</u>: occurs because of increased intake

Clinical features:

-Growth retardation, anaemia, diarrhea

-Copper, cysteine is effective in removing molybdenum from body and decreasing toxicity.</td></tr>
<tr><td>Chromium (Cr)</td><td><u>Function</u>

Its role in human nutrition is not certain.

1. Functions primarily as component of glucose tolerance factor and chief symptoms of chromium deficiency is impaired glucose tolerance test. It prolongs action of insulin.

2. Prevents dental caries.

3. Required for bone development.

4. It may role in lipoprotein metabolism, gene expression.</td><td><u>Disorder of Chromium</u>

<u>Toxicity:</u>

Occupational exposure to chromium in tanning industry:

Renal and hepatic failure

Lung damage, cancer

Tobacco is rich in chromium

Stainless steel utensils also provide chromium to food.</td></tr>
</table>

Nickel (Ni)	Component of enzymes: urease, arginase, acetyl CoA synthetase contain nickel. Required for growth and reproduction.	*Disorder of Nickle* *Deficiency symptoms*: unknown *Toxicity*: non toxic Occupation hazard: respiratory and dermatitis.

INHERITED METABOLIC DISORDERS

Inborn Errors of Metabolism		**Features**	
Caused by genetically determined abnormalities that lead to a block in one of many interrelated biochemical pathways. The block frequently leads to accumulation of substrate and secondary metabolites or to deficiencies of downstream products. IEMs are individually rare but collectively many in numbers account for a significant proportion of neonatal and childhood morbidity and mortality.		1. IMD are categorized by type of metabolic pathways that are impaired, such as disorders of carbohydrates, fats or amino acids. 2. These disorders may involve pathways of specific cellular organelles, including lysosomes, mitochondria, and peroxisomes.	
Metabolic disorder can be categorized into:	1. Disorder that give rise to intoxication – accumulation of toxic compounds occurs proximal to metabolic block.	2. Disorder affecting energy metabolism – two types: Mitochondrial energy defects and cytoplasmic energy defects. Mitochondrial energy defects-	3. Disorder involving complex molecules- It involves cellular organelles and includes mainly lysosomal, peroxisomal, glycosylation

		more severe and are generally untreatable while cytoplasmic energy defects are generally less severe	and cholesterol synthesis defects.
Disorder that give Rise to Intoxication	***Energy Metabolism Disorders***		***Disorder Involving Complex Molecules***
	Mitochondrial Energy Defects	***Cytoplasmic Energy Defects***	
Inborn error of *amino acid metabolism* (Phenylketonuria, Maple syrup urine disease, homocystinuria, tyrosinemia)	*Congenital lactacidemi*a (defects of pyruvate transporter, pyruvate carboxylase, pyruvate dehydrogenase and Krebs cycle enzymes)	Glycolysis, glycogen metabolism and gluconeogenesis	Lysosomal storage disorders
Organic aciduria (methylmalonic, propionic, isovaleric)	*Mitochondrial disorder*s – pyruvate dehydrogenase complex deficiency, pyruvate carboxylase deficiency, myoclonic epilepsy with ragged	Hyperinsulinemias	Peroxisomal disorders

	red fibers, mitochondrial encephalopathy with lactic acidosis and stroke, phosphoenolpyruvate carboxykinase deficiency, Leber's hereditary optic atrophy, neuropathy ataxia and retinitis pigmentosa (NARP) and defect of synthesis of coenzyme Q10)		
Congenital urea cycle defects- Citrullinemia, Argininemia, Argino-Succinic Aciduria, Carbamoyl Phosphate Synthase Deficiency, Ornithine transcarbamylase deficiency	*Fatty acid oxidation defects* carnitine palmitoyl transferase I, II deficiency, short-chain acyl-CoA dehydrogenase deficiency, medium-chain acyl-CoA dehydrogenase deficiency, very long chain acyl CoA dehydrogenase deficiency, long-chain 3-hydroxyacyl-CoA dehydrogenase	Creatine metabolism defects	Disorders of intracellular trafficking and processing (*e.g.* alpha-1-antitrypsin, congenital disorders of glycosylation and inborn errors of cholesterol synthesis.

	deficiency, glutaric academia Type II, carnitine uptake deficiency, hydroxymethylglutaryl CoA lyase		
I*ntolerance disorder* (galactosaemia, hereditary fructose-intolerance)	Ketone body metabolism defects	Pentose phosphate pathway	Inborn errors of cholesterol synthesis
Metal intoxications (Wilson disease, Menkes disorder, hemochromatosis) and porphyria	-	-	-

BIOCHEMICAL BASIS OF RESULTING DISEASE

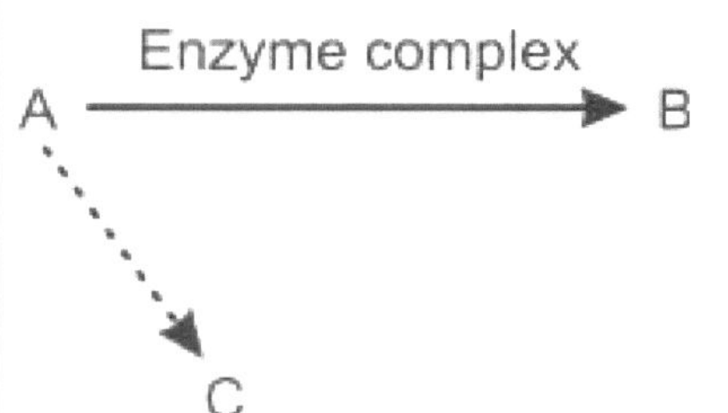

1. Accumulation of a substrate to toxic levels (↑A)
2. Deficiency (due to lack of production) or excessive loss of desired compound (↓B)
3. Conversion of accumulated compound to altered/toxic metabolite (↑C)

Fig. Basis of IMD

Inherited metabolic diseases are the result of gene mutations causing production of an abnormal protein or preventing synthesis of a protein.

Clinical Manifestations: IEM

- IEMs can affect any organ system and symptoms ranges from those of acute life-threatening disease to subacute progressive degenerative disorder or as chronic and progressive neurological symptoms.
- ***In Neonates and Children***, manifestations are nonspecific and mainly remains unnoticed.
  - There may be dysmorphic features present at birth (generally when fetal energy is affected) or develop during the first year of life (lysosomal disorders).
  - In newborns *can present as acute illness* such as encephalopathy - lethargy, irritability, coma, vomiting, respiratory distress, exercise intolerance, cardiac, renal, liver or other organ failure.
  - *In adults*, the symptoms can be *chronic and progressive* neurological symptoms include neurological symptoms such as developmental delay, mild-to-profound mental retardation, autism, learning disorders, behavioral disturbances, muscle weakness, progressive paraparesis, hemiparesis, dystonia, chorea, ataxia, ophthalmoplegia, visual deficit, epilepsy, hepatosplenomegaly and hypoglycemia.

Newborn Screening	***Aim***	To identify infants with serious but treatable disorders so as to prevent or ameliorate the clinical consequences of the disease. Screening is done on blood spots dried on filter paper which is usually obtained by heel-stick

Phenylketonuria	***Urea Cycle Disorders***	***Maple Syrup Urine Disease (MSUD)***
The original bacterial inhibition assays have been abandoned. Now more accurate, faster and efficient methods are available. Alternative methods are available: *colorimetric, fluorometery and MS/MS*	*Citrullinemia*- assessed by measuring citrulline. *Argininosuccidinic aciduria*- assessed by measuring arginine levels. *Ornithine transcarbamylase (OTC) deficiency*- detection of pyroglutamic acid (derived from	Second most common aminoacidopathy. Detected by chromatography, bacterial inhibition assay of leucine and valine, MS/MS detects peak of isoleucine, and hydroxyproline.

(estimate phenylalanine to tyrosine ratio).	glutamine) & glutamine- detected by blood spot assay. *Hyperornithinemia, hyperammonemia, homocitrulinemia and ornithine aminotransferase deficiency*- Ornithine levels assessed.	
Cystathionine Beta Syntheses Deficiency Diagnosed by elevated methionine levels. high methionine and phenylalanine ratio.	***Non-Ketotic Hyperglycemia (NKH)*** Glycine levels are elevated.	***Galactosemia*** Quantitative assay for galactose and galactose-1-phosphate. Confirmatory assay using galactose-1-phosphate uridyl transferase.
Organic Acid Disorder Methylmalonic, propionic and isovaleric academia are characterized by elevated levels of respective amino acids. In glutaric acidemias (both type I and type II)- elevated levels of glutarylcarnitine. Biotidinase deficiency- detected by specific enzyme assay on dried blood spot.	***Fatty Acid Oxidation Disorder*** Carnitine uptake disorders and mitochondrial beta oxidation defects - Acyl carnitine measured by MS/MS. Medium chain acyl CoA dehydrogenase deficiency- elevated octanoylcarnitine (C8), decanoyl carnitine (ratio of C8/C10) and acetyl carnitine (ratio of C8/C2). Confirmed diagnosis – Elevated urinary acylglycines and abnormal plasma acylcarnitine. Definitive diagnosis – Enzymatic or DNA analysis or acylcarnitine profiling in cultured skin fibroblasts. Very long chain acyl CoA dehydrogenase deficiency- abnormal spot test followed by confirmatory genetic analysis or enzyme assay.	

	Short chain acyl CoA dehydrogenase deficiency- often asymptomatic.
Glycogen Storage Disorders All GSD are inherited as AR trait except two both of which are X-linked diseases: Phosphoglycerate kinase deficiency (a muscle glycogenosis) One form of *phosphorylase kinase defici*ency (a hepatic glycogenosis) *Most are childhood disorders*, only few (like type V, McArdle disease) are adult disorders.	GSD are classified according to organ involved and two main categories are born: • GSD involving principally the liver: hepatic glycogenosis • GSD principally involving the muscles: muscle glycogenosis.
HEPATIC GLYCOGENOSIS- presents with hypoglycemia and hepatomegaly	
With hypoglycemia	*With hepatomegaly*
GSD-I= glucose-6-phosphatase deficiency	GSD VI= phosphorylase deficiency
GSD-III= debranching enzyme deficiency	GSD IX= phosphorylase kinase deficiency
GSD 0= hepatic glycogen syntheses deficiency	GSD IV= hepatic branching enzyme deficiency
GSD XI= Glut 2 deficiency	
Muscle Glycogenosis/Cardiac Glycogenosis ***<u>Includes:</u>*** GSD V (muscle phosphorylase deficiency) GSD VII (phosphofructokinase deficiency), GSD X, GSD XII,	***GSD Presenting with Myopathy/Cardiomyopathy*** are: Type II a (lysosomal acid Maltese deficiency) Type II b (lysosomal associated membrane protein- 2)

GSD XIII	
Screening For Inborn Errors ***Phenylketonuria Screening Tests***	
Ferric chloride test: Take 5 ml of fresh urine sample, add 3-4 drops of ferric chloride solution – phenylpyruvate gives green/blue colour	Dinitrophenylhydrazine (DNPH) test: Take 2 ml of urine and add an equal amount of DNPH reagent and mix - wait for 10 mins: appearance of yellow precipitate implies the presence of keto acids: phenylpyruvate. DNPH + Phenylpyruvate = Hydrazine (Yellow ppt)
***Alkaptonuria**:* Deficiency of homogentisate oxidase - darkening of urine on standing: Alkaptonuria (Screening tests)	a. Benedict's reagent – brown colour b. Ferric chloride test – immediate black colour c. Saturated silver nitrate test – immediate black colour
Maple Syrup Urine Disease- Deficiency of the enzyme – branched chain α-keto acid dehydrogenase complex leads to block in the metabolism of branched chain amino acids (valine, leucine, isoleucine). Plasma and urinary levels of valine, leucine, isoleucine, α-keto acids and α-hydroxy acids are raised. Smell of maple syrup (burnt sugar) occurs in urine due to keto- acids	***Maple Syrup Urine Disease (Screening Tests)*** Guthrie's bacterial inhibition test: 4-Azaleucine antagonist of L-leucine Ferric chloride test – greenish grey colour DNPH test is positive

Genetic Screening	Genetic screening plays a major role in diagnosis and population study of genetic disease.
Test	*Condition*
Buccal smear (Barr bodies)	Abnormality of X chromosomes

<table>
<tr><td>Serum assay: Hexosaminidase</td><td colspan="2">Tay-Sachs disease carriers</td></tr>
<tr><td>Phenylalanine levels</td><td colspan="2">PKU screening in newborn</td></tr>
<tr><td>Maternal α-fetoprotein</td><td colspan="2">Neural tube defect, Down's syndrome</td></tr>
<tr><td>Hb by HPLC</td><td colspan="2">Thalassemia carriers</td></tr>
<tr><td>Stool trypsin</td><td colspan="2">Cystic fibrosis</td></tr>
<tr><td>Serum CK</td><td colspan="2">Duchenne muscular dystrophy</td></tr>
<tr><td>Serum methionine</td><td colspan="2">Homocystinuria</td></tr>
<tr><td colspan="3">Lab Investigations Required in IMD</td></tr>
<tr><td colspan="2">Hematology - full blood count (Neutropenia and thrombocytopenia seen in propionic and methylmalonic acidemia)</td><td rowspan="2">Clinical Chemistry Lab Tests
• Electrolytes: Na, K, Cl, HCO_3, blood gases
• Arterial blood lactate (normal values 0.5-1.6 mmol/L)
• Plasma ammonia (normal values in newborn: 90-150µg/dL or 64-107 µmol/L)
• Renal function test: urea, creatinine, NH_3
• Liver function test
• Thyroid function test
• Lipid profile
• Glucose, ketone bodies (in urine)</td></tr>
<tr><td colspan="2">Special Tests
Quantitative amino acid analysis: diagnosis of organic academia and aminoacidopathies
Urine: Orotic acid, organic aciduria
Metabolites by MS (mass spectrometry)
Blood: Specific enzyme assay
Leukocyte, fibroblasts: Enzyme assays</td></tr>
</table>

	• *Others: Uric acid, porphyrins*
Neuroimaging Maple syrup urine disease (MSUD)- brainstem and cerebellar edema Propionic and methylmalonic academia- basal ganglia signal change Glutaric aciduria- frontotemporal atrophy, subdural hematoma	*Electroencephalography* - Mb like rhythm in MSUD, burst suppression in NKH, holocarboxylase synthetase deficiency CSF amino acid analysis- CSF glycine levels elevated in NKH DNA analysis Histology of affected tissue

Clinical Laboratory Assessment of Inherited Diseases		
1.	Laboratory diagnosis:	Clinical features Assessment of metabolite accumulation Absence of enzyme activity Increased concentration of precursor protein Lack of product Biopsy of affected organ Enzyme assay in RBC, WBC, *etc*.
2.	Genetic testing:	DNA-based assays: Techniques: DNA chip DNA sensors Microfluidic DNA sensors Use: Direct detection of mutant genes, *e.g.* cystic fibrosis

		Duchenne muscular dystrophy
3.	Neonatal screening:	Detection of PKU Congenital hypothyroidism
4.	Prenatal diagnosis:	Techniques: Maternal plasma screening: α-fetoprotein Down syndrome Organic acidemias Amniocentesis Chorionic villus sampling Ultrasonography Fetoscopy Cordocentesis Fetal sampling: Fetal blood Skin biopsy

Treatment if IEMs: Followings approaches are used for management of IEMs:	
1. *To prevent Catabolism* Administration of calories is used in acute illness of IEMs to slow down the catabolism.	**2. *Dietary Treatment*** Dietary modification is the main treatment modality in some IEMs. It reduces the formation of toxic metabolites by decreasing substrate availability

	Disorder	*Diet*
1	PKU	Low phenylalanine diet
2	MSUD	Diet low in branched chain amino acids
3	Urea cycle disorders	Protein restriction
4	Organic acidurias	Protein restriction
5	Pyruvate dehydrogenase deficiency	Ketogenic diet
6	Glucose transporter 1 deficiency	Ketogenic diet
7	Galactosemia	Galactose restriction
8	Fructose intolerance	Fructose restriction

3. *Enzyme Replacement Therapy*

This can be used in treatment of some lysosomal storage disorders.

	Disorders	***Enzyme***
1	Gaucher disease	β-glucosidase
2	Pompe's disease	Human alpha glucosidase enzyme
3	Fabry's disease	Recombinant alpha-Gal A
4	Mucopolysaccharidoses I	L- iduronidase

4. *Cofactor Replacement Therapy*

Various enzyme depends on cofactors such as vitamins and minerals for their catalytic properties.

Role of Cofactors in Treatment of Inborn Errors of Metabolism

	Inborn Errors of Metabolism	*Cofactor*

1	Mitochondrial disorders Thiamine responsive variants of Maple syrup urine disease Pyruvate dehydrogenase deficiency/Leigh's disease	Thiamine
2	Glutaric aciduria Type I and Type II Respiratory chain disorders	Riboflavin
3	50% of cases of homocystinuria due to cystathionine β-synthetase deficiency Pyridoxine dependency with seizures Xanthurenic aciduria Primary hyperoxaluria type I Hyperornithinemia with gyrate atrophy	Pyridoxine
4	Methylmalonic academia Homocystinuria	Cobalamin
5	Hereditary orotic aciduria Methionine synthase deficiency Cerebral folate transporter deficiency Hereditary folate malabsorption Kearns-Sayre syndrome	Folinic acid
6	Biotidinase deficiency Propionic academia	Biotin

	Holocarboxylase synthetase deficiency	
7	Hartnup disease	Nicotinic acid

5. *Increase Excretion of Toxic Metabolites*

Toxic metabolites can be removed from the body various modalities such as exchange transfusion, dialysis, forced diuresis.

Carnitine is useful in elimination of organic acids in the form of carnitine esters, sodium benzoate and phenylacetate are also useful in treating hyperammonemia *etc*.

6. Replacement of the end Product

Glycogen storage disorders mostly presents with hypoglycemia which can be prevented by frequent feeds. Raw cornstarch (2 g/kg every 6 hours) can be given in these patients to prevent hypoglycemia, decreasing hyperlipidemia, hyperuricemia and lactic acidemia.

7. Transplantation and Gene Therapy

Many lysosomal storage diseases and peroxisomal disorders can be treated by hematopoietic cell transplantation (HCT).

The main rational for HCT in IEMs is based on the provision of correcting enzymes by donor cells within and outside the blood compartment.

8. *Supportive Care*

Treatment of seizures (avoid sodium valproate – may increase ammonia levels)

Maintain euglycemia and normothermia

Fluid, electrolyte & acid-base balance,

Treatment of infection

Mechanical ventilation if required.

XENOBIOTICS

<table>
<tr><td>Gk. Xenos- stranger; Xenobiotics are chemical compounds foreign to the body. Examples:
• Drugs
• Food additives
• Chemical carcinogens
• Environmental pollutants</td><td>Fact File
At least thirty different enzymes catalyze reactions of xenobiotic metabolism.
Liver is main organ involved in their metabolism.
Other organs active in xenobiotic metabolism are kidney, GIT, gonads, adrenals, placenta and nasal epithelium.</td></tr>
</table>

Metabolism of Xenobiotics: Two Phases

Phase I	***Phase II***
1. *Major Reaction*: hydroxylation Other reactions: deamination, epoxidation, peroxygenation, reduction, nitro reduction, azo reduction. 2. *Enzymes:* *i. Main Enzyme*: monooxygenases or cyt P450; *ii. Other enzymes:* • Flavin-contain mono oxygenase (FMO) • PG synthetase • Alcohol dehydrogenase	1. *Major Reaction*: Conjugation: Conjugation occurs with: glucuronate, sulfate, acetate, glutathione, amino acids, methylation 2. *Outcome*: i. Hydroxylated compounds of phase I are converted to polar metabolites. ii. This increased water solubility facilitates their excretion from body 3. *Mechanism:* i. Phase II metabolism adds an endogenous substrate to the molecule. ii. Results in large increase in water solubility, thus easily excreted

• Epoxide hydrolase • Glutathione reductase • MAO iii. Limited number of enzymes with broad substrate specificity.	4. *Location:* mainly in cytosol ***Phase II Reaction Details*** *1. Glucuronidation* • Most frequent
3. Hydroxylation reaction can result in: • Termination of action of drug, or • Conversion of inactive to biologically active compounds, *i.e.*, pro drugs or pro carcinogens are activated. 4. *Mechanism:* i. Phase I metabolism adds or exposes a functional group on the molecule ii. Results in small increase in water solubility 5. *Location:* mainly in ER	• *Glucuronyl donor*: UDP – glucuronic acid • *Enzyme*: glucuronsyl transferase • *Location:* Present in ER, cytosol • *Compounds glucuronidated*: 2 acetyl aminofluorene, phenol, steroids, morphine • *Mechanism*: attachment of glucuronide to oxygen, nitrogen or sulfur of substrates. *2. Sulfation:* *Sulfur donor:* ○ PAPS ○ Adenosine 3'-phosphate 5'- phosphosulfate (active sulfur)
Cytochrome P450 • Named so due to characteristics 450 nm peak in their absorption spectra (when reacted in their ferrous state with CO) • >100 isoforms, 18 families, 41 sub families	*Compounds Sulfated*: ○ Alcohol ○ Acyl amines ○ Phenols ○ Steroids ○ Glycosamainoglycans

• Carry out metabolism of exogenous and endogenous (*e.g.* steroids) compounds • They are mono oxygenase Reaction catalyzed by cyt P450: $RH + O_2 + NADPH + H+ \rightarrow R - OH + H2O + NADP$	○ Glycolipid ○ Glycoprotein

Fact file: Cytochrome P450
• Electron- transport heme proteins. Heme iron acts as catalytic centre
• Located in smooth endoplasmic reticulum, mitochondria
• Liver contains highest amounts, also found in small intestine, lung, and brain.
• NADPH involved in reaction mechanism of cyt P450, NADH not involved.
• Not involved in phase II.
Factors influencing P450 metabolism:
- Specific difference
- Individual difference
  • Genetic make up
  • Age
  • Gender
	3. Glutathione (GSH)
	$R + GSH \rightarrow R- S- G$
	• Detoxifies electrophilic xenobiotics.
	• Catalyzed by glutathione S-transferase
	GSH- fact file:
	• Tripeptide
	• Gamma-glutamyl cysteinyl glycine
	• Glutamic acid, cysteine, glycine
	• Contains sulfhydryl groups of cysteine
	• Glutathione conjugates get further metabolized before excretion
	• Glutamyl and glycinyl groups of GSH are removed and acetyl group (from acetyl CoA) gets attached to cysteinyl moiety to form mercapturic acid which is then excreted.

<table>
<tr>
<td>

- Fitness/ disease
- Diet

- Drug- drug and xenobiotic- drug interactions

Nomenclature of cytochrome *P450*

- Abbreviated as CYP
- Followed by Arabic number for family
- Followed by capital letter for subfamily
- Then, individual P450 are assigned Arabic numerals.

</td>
<td>

- **Functions of GSH**
  - Xenobiotic metabolism
  - Decomposition of hydrogen peroxide
  - *Important intracellular reductant*: Maintains SH group of enzymes in reduced state
  - *Transport of amino acid across membrane in kidney:*

Amino acid + GSH –GGT→ Glutamyl amino acid + cysteinyl glycine

Example: conjugation of glutathione to cyclophosphamide

</td>
</tr>
<tr>
<td>

- CYP1A1: Cyt P450 that is member of family 1, sub family A, first individual member of this subfamily.
- Nomenclature for gene is same as of CYP with addition of italics.

Inhibitors of CYP

Competitive inhibition *e.g.* Omeprazole: CYP2C19

</td>
<td>

4. Acetylation

- *Donor*: Acetyl CoA (active acetate)
- *Enzyme*: Acetyl transferase
- *Example*: *Isoniazid:* antitubercular drug is acetylated by acetyl transferase.
- Two polymorphic forms of enzyme exist:
- fast, slow acetylators.
- *Slow acetylators* subjected to toxic effect of drug because it persists longer.

</td>
</tr>
<tr>
<td>

Clinical implication of drug interaction due to induction of CYP

Occurs when effect of one drug are altered by prior, concurrent or later administration of

</td>
<td>

5. Methylation

Donor: Adenosyl methionine

Enzyme: Methyl transferase

</td>
</tr>
</table>

another. *E.g.,*	
1. Anticoagulant Warfarin: It is metabolized by CYP2C9. If patient is given phenobarbital concomitantly, the dosage of warfarin must be changed. Phenobarbital induces CYP2C9 and warfarin is metabolized more quickly than before and dosage becomes inadequate increasing risk of bleeding in such patient.	*2. Ethanol Consumption and Smoking* Ethanol induces CYP2E1. Tobacco smoke components are metabolized by cyt P450 and many of these components are carcinogenic. Elevation of CYP2E1 activity by alcohol may increase risk of carcinogenicity from exposure to compounds of tobacco smoke.
3.PAH metabolism: CYP1A1 metabolizes PAH (polycyclic aromatic hydrocarbons). PAH (procarcinogen) is inhaled by smoking and gets converted to active carcinogen by CYP1A1. Also, PAH can cross placenta if women is smoking.	*4.CYP polymorphism:* Certain cyt P450 exist in polymorphic forms (genetic isoforms), some of which have low catalytic activity. This is responsible for variation in drug response among many patients.
Therapeutic Inhibition of CYP Without an understanding of drug metabolism and drug interaction mediated by CYP, medication cannot be safely prescribed.	
CYT P450 Drug Interactions	

Cyt	*Drug*	*Effect*
CYP2E1	Alcohol	Induces CYP
	Isoniazid	Induces CYP

	Halothane	Metabolized to compounds that damage liver protein. *Outcome*: People with high CYP2E1 activity due to alcohol intake are at higher risk of hepatitis reaction to an anaesthetic.
CYP3A4	Rifampicin	Induce it Taking Rifampicin with oral contraceptives (metabolized by CYP3A4) lower efficiency of contraceptive
	Carbamazepine	Induces it
	Phenytoin	
	Glucocorticoid	
	Ketoconazole	Ketoconazole with warfarin leads to excessive bleeding.
	Cyclosporins	
	Erythromycin	Induces it

Genetically Determined Defects in Drug Reactions		**Effects of Xenobiotics**
Enzyme Affected	***Consequence***	Diverse effects are produced by drugs/ xenobiotics: • Inactivation, activation, formation of a toxic metabolite • Inactivation of active form of drug to inactive form
G6PD	Hemolytic anaemia following primaquin intake	
Ca^{2+} release channel	Malignant hyperthermia	
CYP2D6	Slow metabolism of drugs	

<table>
<tr><td>CYP2A6</td><td>Impaired metabolism of nicotine</td><td>decrease bioavailability of drug</td></tr>
<tr><td colspan="3">

- *Toxicity:* Activation of drug by cyt P450 to biologically active form a toxic form
- *Cell injury* (cytotoxicity) can be severe to cause cell death
- *Reactive species of xenobiotics* produced by metabolism bind to cell macromolecules covalently that can be DNA, RNA, proteins.
    - Proteins: alter its antigenicity: Xenobiotic acts as a haptan and result in antibody production.
    - Binding to DNA: causes chemical carcinogenesis.

</td></tr>
</table>

<table>
<tr><td>

Indirect Carcinogens

Certain chemicals require activation by monooxygenases in ER to become carcinogenic.

Other chemicals interact directly with DNA*: DIRECT CARCINOGENESIS*

</td><td>

Epoxides

- Products of action of certain monooxygenases
- Highly reactive, mutagenic or carcinogenic

Epoxide hydrolase acts on them to convert them to less reactive dihydrodiols.

</td></tr>
<tr><td colspan="2">

Role of Xenobiotics in Disease

</td></tr>
<tr><td>

- In human, extensive molecular mechanisms exist for protection from toxic exogenous xenobiotic compounds.
- Many detoxification enzymes namely, cytochrome P450s and flavin-containing monooxygenases have now been reported to have endogenous roles in the *pathogenesis of metabolic diseases*.
- Exposure to environmental pollutants can perturb the physiological processes and result in *increased risk of developing*

</td><td>

- Many xenobiotics are able to induce fatty liver.
- Xenobiotics *can cross placenta freely* and environment pollutants can cause damage to growing fetus causing *congenital anomalies*.
- Data indicate that xenobiotic- metabolizing enzymes involved in metabolism of endogenously formed or diet derived compounds that are themselves associated with disease pathologies.

</td></tr>
</table>

chronic diseases such as cardiovascular disease and diabetes.	

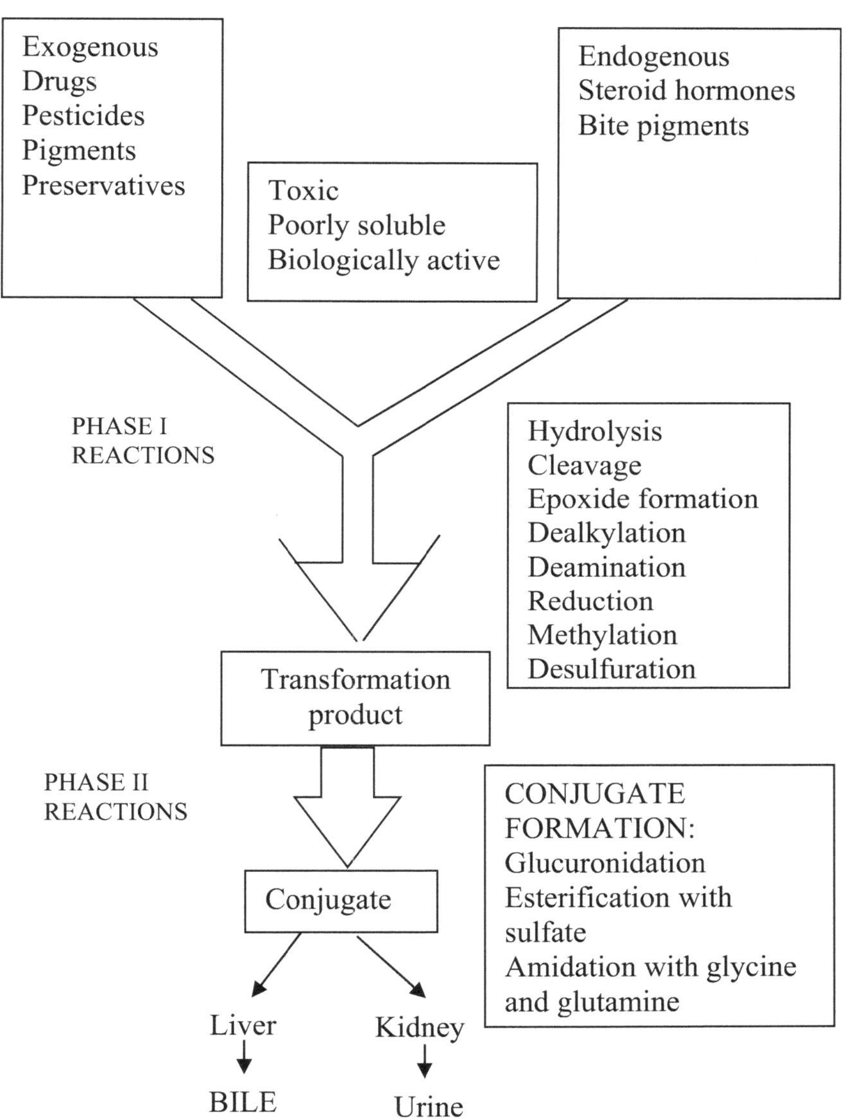

OXYGEN: OXIDATIVE STRESS AND ANTIOXIDANT DEFENCES

Free Radicals	**Biomedical Importance**
Any species capable of independent existence that contains one or more unpaired electrons. Radicals can be formed by loss of a single electron from a non-radical, or by gain of a single electron by a non-radical or breakage of covalent bond 'homolytic fission.'	• Free radicals are formed in the body under normal conditions • They cause damage to nucleic acids, proteins, and lipids in cell membranes and plasma lipoproteins • This can causes cancer, atherosclerosis and coronary artery disease and autoimmune disease • Epidemiological studies have shown protective role of antioxidant nutrients namely, selenium, vitamin C and E, β carotene, polyphenol. • International trials show little benefit of antioxidant supplements.
Free Radical Reactions	They are self-perpetuating chain reactions as they are highly reactive molecular species with an unpaired electron. They have a very short time (10^{-9} – 10^{-12} sec) and either abstract or donate an electron in order to generate a new radical from the molecule with which they colloid.

Oxygen radicals are the most damaging radical in biological system

ROS (Reactive oxygen species)		**Role of Free Radicals in Biologic Systems**
Radicals	*Non Radicals*	Enzyme – catalyzed reactions
Super oxide $O2^{\bullet}$	H_2O_2 (hydrogen peroxide)	Electron transport system in mitochondria
Hydroxyl $HO^{\bullet}$	HOCl (hypochlorous acid)	Signal transduction and gene expression
Peroxyl $ROO^{\bullet}$	Ozone (O_3)	Activation of nuclear transcription factors
Alkoxyl $RO^{\bullet}$	Singlet oxygen $(O)^{-}$	Oxidative damage of molecules cells, tissues

<table>
<tr><td colspan="2">Hydroperoxyl $HOO^{\bullet}$ Peroxynitrite ($ONOO^-$)</td><td>Antimicrobial actions

Aging and disease</td></tr>
<tr><td colspan="3">Sources of ROS (Fig. 14.2)</td></tr>
<tr><td>Mitochondrial and microsomal ETC</td><td>In peroxisome: FA oxidation generates H_2O_2</td><td rowspan="2">Cyt P_{450} monoxygenase:
• In detoxification of drugs e^- transported from NADPH to O_2.
• Leakage of electrons during these conversions</td></tr>
<tr><td>NADPH oxidase: Enzyme is present in phagolysosome, in immune cells, e^- transfer from O_2 to form O^{2-}.</td><td>Respiratory burst: Myeloperoxidase (MPO) in neutrophils catalyze formation of HOCl from H_2O_2</td></tr>
<tr><td>Ionizing radiation (X-rays, UV)</td><td colspan="2" rowspan="2">• Fenton reaction: A mixture of hydrogen peroxide and iron (II) salt causes formation of hydroxyl radical.

$O^{2-} + H_2O_2 \xrightarrow{\text{metal}} O_2 + HO^{\bullet} + OH^-$</td></tr>
<tr><td>Transition metal ions, including Cu^{+2}, Co^{+2}, Ni^{+2}, Fe^{2+}</td></tr>
<tr><td>Enzymatic reduction of oxygen:</td><td colspan="2">$\text{Xanthine/hypoxanthine} \xrightarrow{\text{Xanthine Oxidase}} \text{Uric acid}$
O_2 → O_2^-</td></tr>
<tr><td colspan="3">Nitric oxide (EDRF) itself a radical, also called endothelial derived-relaxing factor (EDRF) and has following functions: Vascular function, platelet aggregation, immune response neurotransmitter, signal transduction, cytotoxicity.</td></tr>
<tr><td colspan="2">Oxidative stress: imbalance between reactive oxygen species (ROS) and antioxidants.
Antioxidant
Prooxidant</td><td>Under normal conditions, cells are able to balance production of oxidants and antioxidants.</td></tr>
</table>

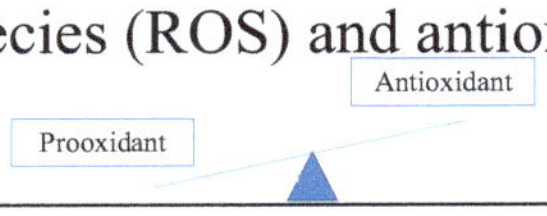

• Imbalance between prooxidant and antioxidant • Damage caused by free radical / reactive oxygen species	Oxidative stress occurs when cells are subjected to excess levels of ROS or as a result of antioxidant depletion. Under normal conditions, ROS are natural byproducts. It also can be generated by exogenous sources. Major sources of ROS: mitochondria, peroxisome, plasma membrane/cytoplasm
Deleterious Activities of Free Radicals and Oxidants and Pathogenesis When produced in excess, free radicals and oxidants generate oxidative stress. **Oxidative Stress** A deleterious process that can seriously alter the cell membranes and other structures such as proteins, lipids, lipoproteins, and deoxyribonucleic acid (DNA). The body has several mechanisms to counteract these attacks by using DNA repair enzymes and/or antioxidants. If not regulated properly, oxidative stress can induce a variety of chronic and degenerative diseases as well as the aging process and some acute pathologies (trauma, stroke).	**Free Radical And Diseases** Cancer Inflammation/ infection Ischemia reperfusion injury: cerebral, cardiac Neurodegenerative diseases Cardiovascular diseases Autoimmune disease Atherosclerosis. Aging Others: *e.g.* drug, chemical – induced toxicity *etc.*

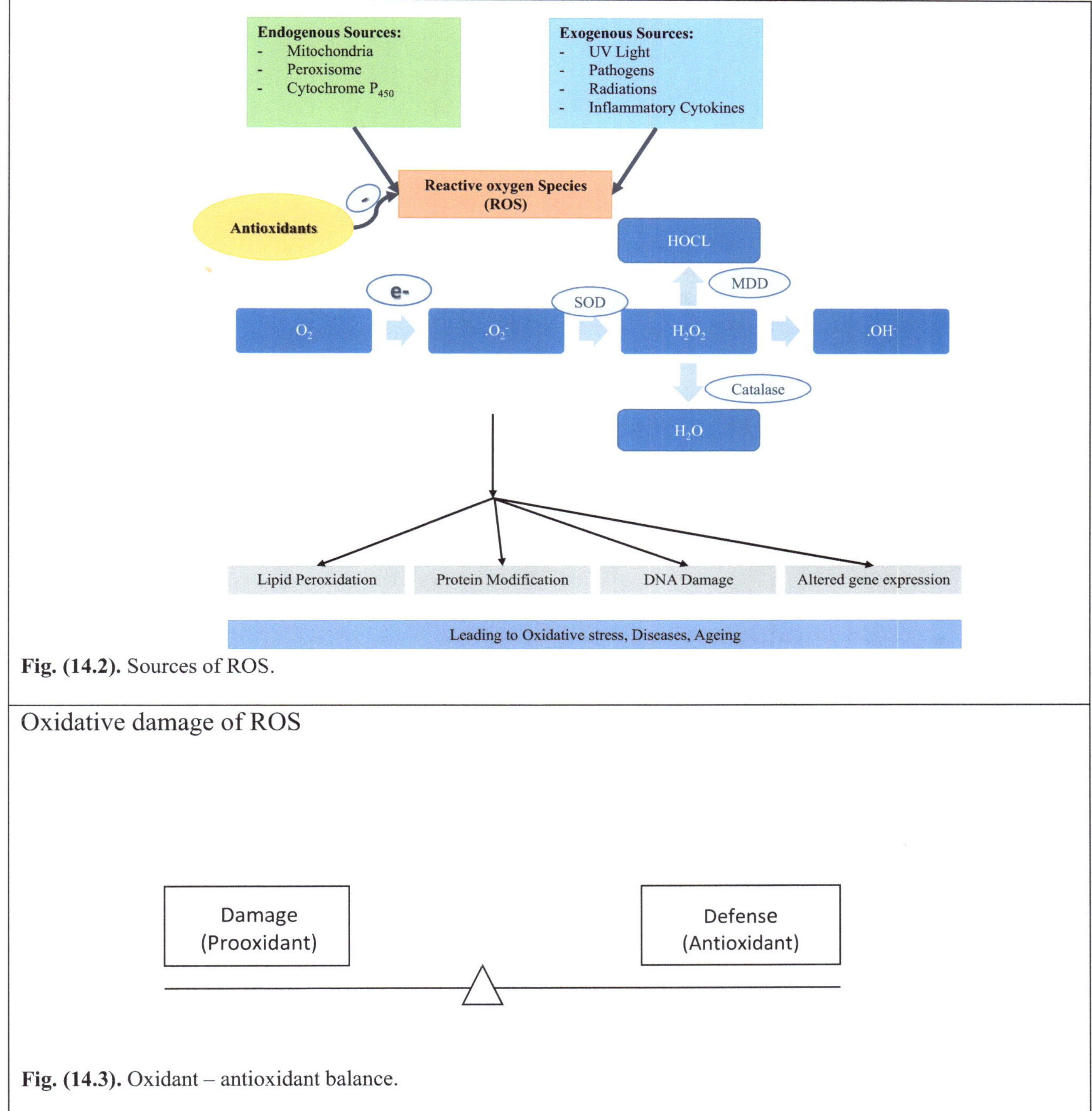

Fig. (14.2). Sources of ROS.

Oxidative damage of ROS

Fig. (14.3). Oxidant – antioxidant balance.

Mechanisms of Protection Against Radical Damage

Antioxidants

Compound that can delay the start or slow the rate of lipid oxidation reaction.

Impaired endogenous antioxidant system results in accumulation of free radicals, which not only induces lipid peroxidation but also imposes severe stress on the body leading to many diseases

Antioxidant Defense System

Endogenous Antioxidant Molecules

- Antioxidant enzymes: SOD (superoxide dismutase), CAT (catalase), GSHPx (Glutathione peroxidase)
- Non-enzymatic antioxidants: GSH, Bilirubin

Exogenous Antioxidant Molecules

- α tocopherol – prevents oxidation of FA
- Carotenoids - β-carotene, lycopene destroy a particularly damaging form of singlet oxygen.
- Ascorbic acid – radical scavenging, recycling of vitamin E
- Riboflavonoids – potent antioxidant activity.

Sequestration of Metals

- Transferrin, lactoferrin, ferritin, metallothionine

Repair Systems

- DNA repair enzymes, metalloproteinases, GST

Superoxide Dismutase (SOD)

$2O^{2-} + 2H^{+} \rightarrow H_2O_2 + O_2$

- Only enzyme known to react with radical
- Presence of SOD implies O^{2-} produced in cell during normal metabolism
- SOD is a primary antioxidant enzyme

GSH cycle (Glutathione Cycle)

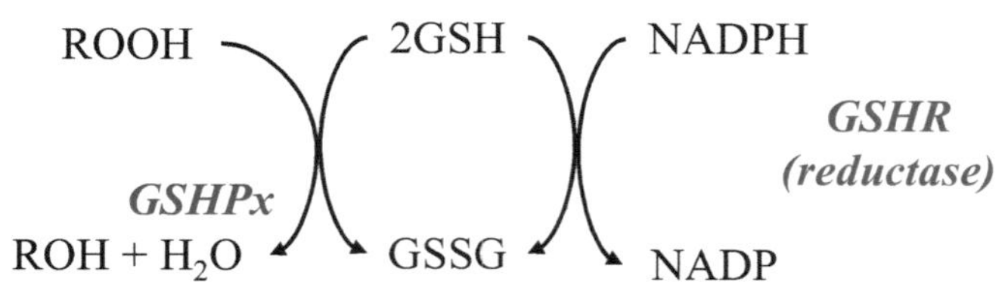

ROOH → ROH + H_2O (*GSHPx*), coupled with 2GSH → GSSG

GSSG → 2GSH (*GSHR (reductase)*), coupled with NADPH → NADP

<table>
<tr><td>Catalase (CAT)

Neutralizes H_2O_2

$2H_2O_2 \rightarrow 2H_2O + O_2$

Prevents lipid peroxidation and protein oxidation</td><td>GSHPx: Function

Removes H_2O_2 and ROOH

$ROOH + 2GSH \rightarrow ROH + H_2O + GSSG$

Deficiency in GSHPx leads to oxidative hemolysis

Protects against lipid peroxidation</td></tr>
<tr><td>Ascorbic Acid

Donate $1e^-$ → semi dehydroascorbate

DHA + 2GSH → ascorbate + GSSG</td><td>Tocopherol

Chain breaking antioxidant:

Scavenging peroxy radical

Inhibits chain reaction of lipid peroxidation</td></tr>
<tr><td colspan="2">Many epidemiological data suggest that antioxidants may have a beneficial effect on many chronic diseases, the systematic use of supplements is hindered by several factors: the lack of prospective and controlled studies, the long-term effects and the dosages necessary for each type of diseases.</td></tr>
</table>

QUESTIONS

1. Define/ Write short notes on: -

- Absorption of iron.
- Anion gap.
- Beri Beri
- Calcium homeostasis
- Copper deficiency
- Dietary fiber
- Fluoride
- Folic acid deficiency
- Gla
- Hemosiderosis
- Hypercalcemia
- Hyperkalemia
- Hyperphosphatemia
- Hypocalcaemia.
- Hyponatremia
- Iron metabolism.

- Kwashiorkor
- Leucine pellagra
- Marasmus
- Night blindness
- Nitrogen Balance
- Nutritional disorders
- Obesity
- Osmolality
- Pellagra
- Pernicious anaemia
- Respiratory quotient
- Riboflavin
- Rickets
- Scruvy
- 1-C metabolism
- Selenium
- Significance BMR
- Special diets
- Tetany
- Transferrin
- Vitamin E
- Vitamin K
- Vit K cycle
- Wald's Visual cycle
- Water intoxication
- Wilson's disease
- Factors effecting absorption of iron
- Beneficial effects of Dietary fiber
- Coenzyme functions of vitamin B
- Vit D deficiency disorders
- Hormones regulating calcium
- Megaloblastic anaemia
- Mutual supplementation
- Vitamin D as a hormone

2. Discuss sources, function and deficiency manifestations of iron.
3. Name a hemeprotein.
4. What is glucose tolerance factor?
5. Calculate energy requirement of
   a. Adult man with accountant profession.
   b. Student
   c. Pregnant women
   d. Lactating (nursing) women
   e. Labourer

6. Differentiate between Marasmus and Kwashiorkor.
7. Write important function of vitamin A
8. Enumerate important functions of vitamin D
9. Explain synergistic action of vitamin E and selenium
10. Write the basis of hemolytic anemia in vitamin E deficiency.
11. Why human cannot synthesize vitamin C?
12. Write functions of vitamin C.
13. Define free radicals? What are different sources of free radicals?
14. Explain the mechanism of DNA, lipid and protein damage by ROS.
15. What are different ways of oxidative stress measurement?
16. Write a note on free radicals and disease.
17. Make a list of antioxidants and explain their mechanism of action.
18. Explain the prooxidant and antioxidant nature of vitamin C.
19. Discuss distribution of water, sodium and potassium in various body compartments.
20. Write in brief on regulation of water and electrolyte balance
21. What are different types of buffers present in your body?
22. How will you evaluate acid-base status?
23. Classify Acid base disorders
24. Write a short note on regulation of acid-base disorders
25. Write short note on CYP450
26. What are phase I and phase II reactions of xenobiotics

27. Write brief note on induction of CYP and clinical implication of CYP induction

28. What is the importance of GSH?

BIBLIOGRAPHY

Bruce Alberts, Alexander Johnson, Julian Lewis, Martin Raff, Keith Roberts, and Peter Walter. Molecular Biology of the Cell. 4th edition. New York: Garland Science; 2002.

Denise R Ferrier. Lippincott illustrated reviews: biochemistry. 7th Edition. Philadelphia Wolters Kluwer; 2017

Donald Voet, Judith G Voet, Charlotte W Pratt. Fundamentals of Biochemistry. 5th Edition. New York: Wiley; 2016.

Geoffrey L Zubay, Dennis E Vance. Principles of biochemistry. Dubuque, Iowa: William C. Brown; 1995.

Geoffrey M. Cooper & Robert E. Hausman. The cell: A molecular approach. 7th Edition. Oxford University Press; 2019.

Jeremy M Berg, Gregory J Jr Gatto, Lubert Stryer, John L Tymoczko. Biochemistry. 9th Edition. New York: Macmillan International Higher Education: WH Freeman; 2019.

Keith Wilson, John M Walker. Principles and techniques of biochemistry and molecular biology. 7th edition. Cambridge: Cambridge University Press; 2017.

Lehninger A, Nelson D, Cox M. Lehninger principles of biochemistry. New York: Worth Publishers; 2000.

Michael A Lieberman, Rick E Ricer. Biochemistry, molecular biology, and genetics. 7th Edition. Philadelphia, Pa Wolters Kluwer; 2020.

Victor W Rodwell, David A Bender, Kathleen M Botham, Peter J Kennelly, P Anthony Weil. Harper's illustrated biochemistry. 31st edition. New York: Mcgraw-Hill Education; 2018.

CHAPTER 15

Clinical Biochemistry, Physiology & Genetic Disorders

LEARNING OBJECTIVES:	Keywords:
• Explain the biochemistry of blood, its functions and associated disorders. • Illustrate the concepts of immunology, constituents and diseases of the immune system. • Describe the biochemical basis of structure and functions of muscle, nerves and eye. • Identify and interpret organ function tests. • Define and diagnose genetic disorders.	Action Potential, Antibody, Antigens, Blood, Clotting factors, Calcium, Genetic disorders, Homeostasis, Immunity, Immunoglobulins, Jaundice, Kidney, Lens, LFT, Mucin, Neurons, Platelets, Sarcomere, Tear, Vaccines.

15.1. BLOOD

Biochemistry of Blood	**Blood Performs Three Major Functions (Fig. 15.2)**		
Composition of Blood (Fig. 15.1)	***Transport***		***Defense***
8% of the body weight (5–6 L) Suspension of cells in carrier fluid: Water (90%) Proteins (7%) Inorganic (1%) Organic (2%) Plasma: 55% Leucocytes and platelets: 2%	Oxygen and carbon dioxide Food molecules (glucose, lipids, amino acids)	Ions (*e.g.*, Na^+, Ca^{2+}, HCO^{3-}) Wastes (*e.g.*, urea) Hormones	Defending the body against infections or other harmful foreign materials. All types of WBCs participate in the body's defence mechanism.
	Homeostatic Functions		
	Heat Water- salt balance Acid-base balance		Osmosis Blood clotting Formation of hormones

Simmi Kharb

Erythrocytes: 43%		
Red Blood Cells (Erythrocytes)	Does not contain nucleus, chromatin Does not contain mitochondria	ATP produced from anaerobic glycolysis mainly and ends in lactate (~90%).

Structure	***Glycolysis in RBCs has the Following Features***
In an adult human, hemoglobin (Hb) molecule comprises of four polypeptides Two alpha (α) chains consisting of 141 amino acids Two beta (β) chains consisting of 146 amino acids Each of these chains is attached to the prosthetic heme group. One Fe-atom is present at the center of each heme group. One O_2 molecule binds to each heme group. It is a reversible reaction.	2,3 BPG will be produced and not 1,3 BPG. 2,3 BPG binds O_2 to hemoglobin: Low concentration of 2,3 BPG will increase affinity hemoglobin (Hb) to O_2 PPP is the main pathway for producing of reductive equivalents NADPH It is accountable for transporting O_2 and CO_2.

Diseases Associated with the Inadequate Synthesis of Hemoglobin Components		
Porphyria	Thalassemia	Iron deficiency anemia

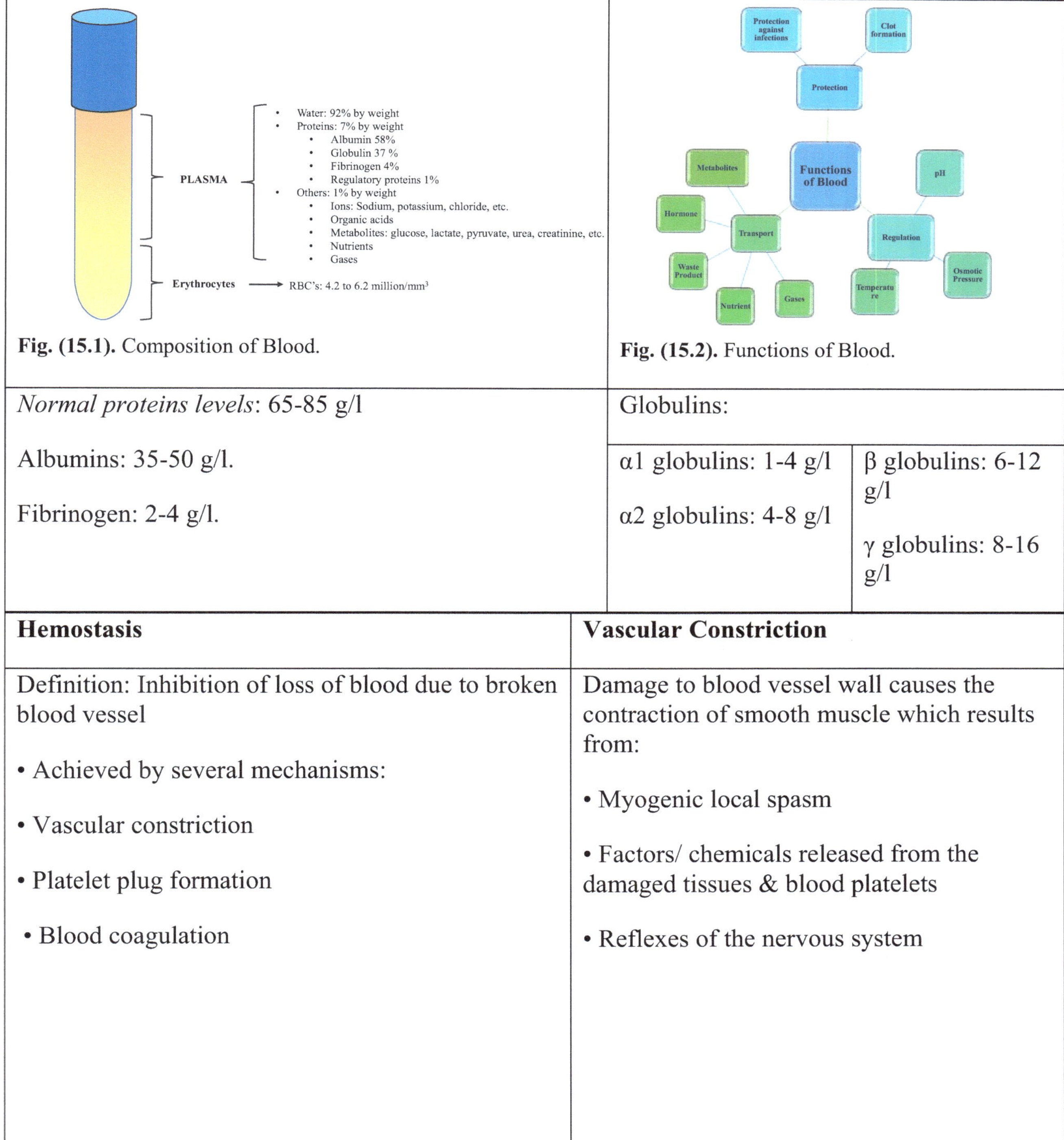

Fig. (15.1). Composition of Blood.

Fig. (15.2). Functions of Blood.

Normal proteins levels: 65-85 g/l

Albumins: 35-50 g/l.

Fibrinogen: 2-4 g/l.

Globulins:

α1 globulins: 1-4 g/l α2 globulins: 4-8 g/l	β globulins: 6-12 g/l γ globulins: 8-16 g/l

Hemostasis

Definition: Inhibition of loss of blood due to broken blood vessel

- Achieved by several mechanisms:
- Vascular constriction
- Platelet plug formation
- Blood coagulation

Vascular Constriction

Damage to blood vessel wall causes the contraction of smooth muscle which results from:

- Myogenic local spasm
- Factors/ chemicals released from the damaged tissues & blood platelets
- Reflexes of the nervous system

Platelets or Thrombocytes

- Smallest formed elements in blood (diameter 1-4 μm)
- Anuclear fragments from megakaryocytes
- Typically about 1000 platelets are produced from a single megakaryocyte
- Development happens in response to the hormone - thrombopoietin

- They may generally be removed from circulation by the action of tissue macrophages
- 150,000-300,000/μl
- Half-life 5 to 12 days

- Constituents in cytoplasm:
- Thrombosthenin, actin & myosin,
- Mitochondria, ER & Golgi apparatus

Dense Granules

- Phospholipids
- Serotonin
- Calcium
- ADP • ATP

α Granules

- von Willebrand factor
- Clotting factor V & Clotting factor XIII
- Platelet derived growth factor (PDGF)

Functions of Platelets

1. Vasoconstriction is caused by serotonin released from platelets
2. Vascular injury gets plugged by aggregation of platelets
3. Clotting factors are provided by the platelets
4. Platelets' contractile protein cause clot retraction
5. Platelets' growth factor induces mitosis in the cells of the vascular wall

Dissolution of Clots

- A clot comprises of a mesh made with fibres of fibrin crossing each other in all directions.

Fibrin fibres entrap blood cells, platelets, and plasma, and also a large quantity of plasminogen in the clot

- Plasminogen is converted to plasmin by tissue plasminogen activator (t-pa)
- Fibrin fibers & other protein coagulants like factors I, II, V, VIII, & XII are digested by plasmin

<table>
<tr><td><u>Blood Coagulation</u></td><td colspan="3">The Activation of Clotting Factors Leads to Blood Coagulation</td></tr>
<tr><td>Three stages of blood coagulation (Fig. 15.3):</td><td>• Formation of prothrombin activator

- Extrinsic pathway

- Intrinsic pathway</td><td>• Conversion of prothrombin to thrombin</td><td>• Conversion of fibrinogen to fibrin</td></tr>
<tr><td colspan="2"><u>Prevention of Blood Clotting: Endothelial Surface Factors</u>

a. Smoothness at the surface of the endothelial cell prevents contact-activation of clotting system (intrinsic)

b. Glycocalyx layer present on the endothelium repels clotting factors & platelets

c. Thrombomodulin binds to thrombin producing thrombomodulin- thrombin complex which activates protein C which in turn inactivates Factors V & VIII</td><td colspan="2">Fibrin

Fibrin fibres adsorb 85 to 90% of thrombin

This prevents spreading of thrombin

The excessive spread of clot Antithrombin III is prevented

Antithrombin III is available to bind with the thrombin not adsorbed to fibrin fibres</td></tr>
<tr><td colspan="2">Bleeding Disorders

• Liver diseases may also weaken the clotting system

• Deficiency of vitamin K may lead to severe bleeding tendencies like Hemophilia

• Deficiency/ abnormality in Factor VIII & IX</td><td colspan="2">• Haemophilia is not observed in women, although he abnormality is present on X chromosome since at least one of the two X chromosomes harbours the appropriate gene

Purpura

• Thrombocytopenia may lead to a group of disease call purpura</td></tr>
</table>

Anticoagulants	**Blood Coagulation Tests**
Anticoagulants like EDTA, sodium citrate, potassium oxalate, *etc*. act by binding with calcium ions Heparin is also a common anticoagulant used both *in vivo* & *in vitro* Dicumarol & Warfarin are Vitamin K antagonists used as anticoagulants *in vivo*	***Bleeding Time***: It is the time from onset of bleeding till the bleeding stops (normal: 1-5 min) ***Clotting Time:*** It is the time from the onset of bleeding till the formation of clot formation (normal: 2-8 min) ***Prothrombin Time:*** It is the time required for the process of coagulation to take place (normal: 12-16 s)

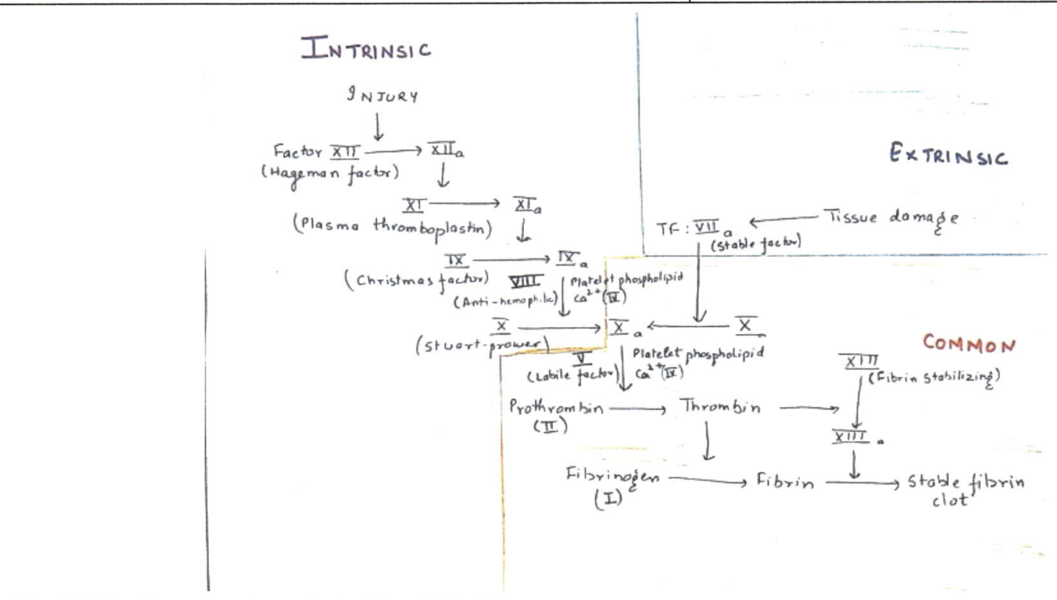

Fig. (15.3). Cascade of events in Blood Clotting.

IMMUNOLOGY

<table>
<tr><th colspan="5">Terminology</th></tr>
<tr><td>Immune System</td><td colspan="4">Unique adaptive defensive system that protects the organism from invading pathogenic microorganisms as well as cancer.</td></tr>
<tr><td>Immunity</td><td colspan="4">Physiological mechanisms present inside an organism that not only recognize foreign as well as neutralize/ metabolize/ eliminate them</td></tr>
<tr><td>Antigen</td><td colspan="4">Any foreign substance having the potential to bind with T cell receptor or antibody.</td></tr>
<tr><td>Antibody</td><td colspan="4">An immunoglobulin protein that can recognize a specific antigen and bind to it precisely.</td></tr>
<tr><td>Innate Immunity or Native Immunity</td><td colspan="4">The resistance possessed by an individual to infections is as per his/her constitutional and genetic make-up.

It is not affected by previous contact with immunization or microorganisms.</td></tr>
<tr><td rowspan="2">Adaptive Immunity or Acquired Immunity</td><td colspan="4">The resistance acquired by an individual through his/her lifetime. It is dissimilar to innate inborn immunity.

Four immunological characteristics are exhibited by adaptive immune response.</td></tr>
<tr><td>Specificity</td><td>Diversity</td><td>Specificity</td><td>Self and non-self recognition.</td></tr>
<tr><td rowspan="3">Immune System</td><td colspan="4">Genes, molecules, cells and tissues constitute the immune system that has ability of recognition and elimination of non self and foreign substances (antigens).

Immune system includes two systems:</td></tr>
<tr><td colspan="2">Cellular immune response.</td><td colspan="2">Cellular immune response.</td></tr>
<tr><td colspan="4">Immune system has two forms:</td></tr>
</table>

	- Innate or nonspecific	- Adaptive or specific

Cellular & Humoral Components of Immune System: Body has two types of immune response (Fig. **15.5**)	
I. Humoral immunity production of soluble antibodies or immunoglobulins in response to immunogen by B -lymphocytes in bone marrow.	II. Cell-mediated response is through cytotoxic or killer T cells.

Immune System: Three Lines of Defense		
First Line of Defense:	***Second Line of Defense:***	***Third Line of Defense:***
Skin, mucus membrane, skin and mucus membranes' secretions	Phagocytic WBCs, antimicrobial proteins, inflammatory response.	Lymphocytes Antibodies

Immune Response	Immune response depends on pathogens and can be categorized as innate or adaptive

Innate Immune System	**Adaptive Immune System**
Non-specific response Maximal immediate response after exposure Components of humoral and cell-mediated immunity Immunological memory not developed. All forms of life (animals & plants) exhibit this immune system	Antigen and pathogen specific response Lag time amongst exposure and maximal response Components of humoral and cell-mediated immunity Immunological memory is developed after exposure Only jawed vertebrates exhibit this immune system.

Innate and Adaptive Immunity	Mediators of inflammatory response: humoral component-

Nonspecific immunity: Body's first line of defense Uses physiochemical barriers (mucosa of epithelium and skin) and their secretions (sweat, mucus, acid) Involves inflammatory response	Polymorphs Mononuclear phagocytes Lymphocytes Platelets	Endothelial cells CRP Complement system

Adaptive or Acquired Immunity	Based on lymphocytic potential to make specific antigenic receptors on encounter with foreign antigens. Lymphocyte carrying receptor for its cognate antigen gets activated to immune response.
Immunoglobulins: Fact File	Produced in response to foreign substances (antigens) Associated with γ-globulin fraction of plasma proteins Diverse unique group of molecules recognize and react with extensive range of the specific antigenic structures. An animal can synthesize many millions of different antibodies, each with ability to interact with specific antigens: *ANTIBODY DIVERSITY*
Antibody diversity occurs due to presence of large number of variable genes, somatic recombination, somatic mutation, deamination of C → U. As lymphocyte mature, each differentiating cell constructs particular L and H genes of virtually unique structure by recombination process.	

Structure Immunoglobulins	Five major classes of immunoglobulin share a common structural form:
Y-shaped molecule contains:	Two identical heavy (H) chain units Two identical light (L) chain smaller units [Ig is a tetramer: H_2 L_2].

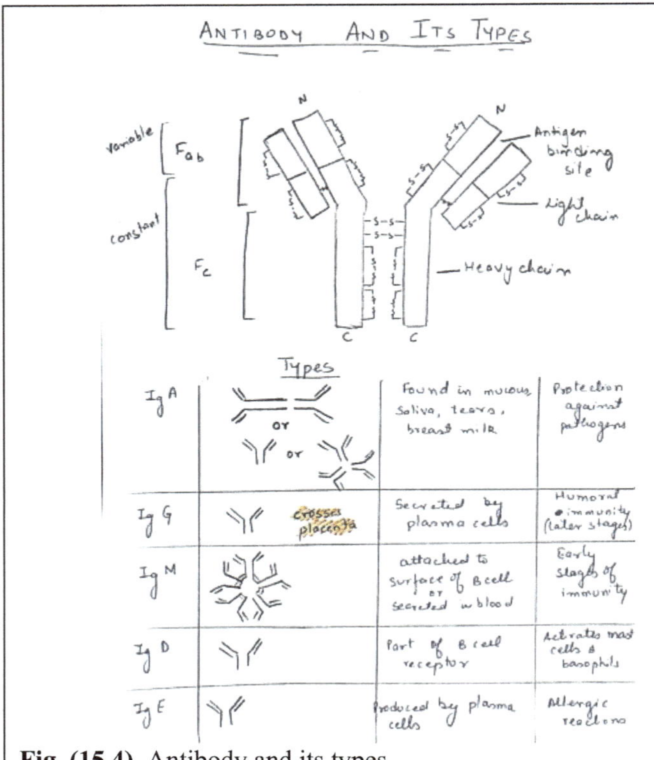

Fig. (15.4). Antibody and its types.

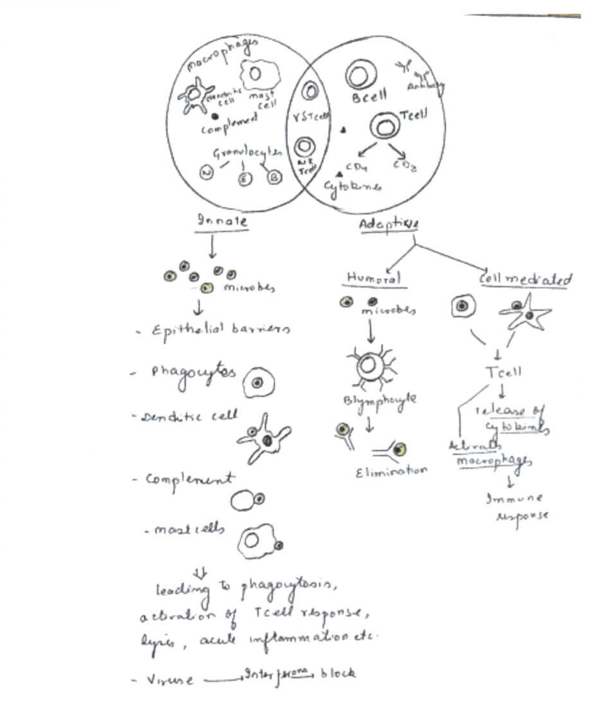

Fig. (15.5). Types and components of Immunity.

Immunoglobulin Class is Determined by Nature of H Chain (Fig. 15.4)				
IgG	IgA	IgD	IgM	IgE
(γ)	(α)	(μ)	(δ)	(∈)
Chains can be of kappa (K) and lambda (γ) type and can be found in any Ig class.				

Antibody Monomer	
Monomer, but composed of two pairs of polypeptide chains Ab monomer made of two identical heavy (H) & two identical light (L)chains.	H and L chains held by disulphide bonds that occur: Between H and L chains Between H chains

Each H chains is paired in same orientation: (-amino to carboxyl within L-chain).	Between L chains
Amino portion of both L and H chains has variable structures that are adjacent to each other forming antibody binding site which binds antigen. Carboxyl terminal of H and L chains have a distinct region which is accountable for Ig function other than recognition of epitope *e.g.*, activation of complement.	H chains, further consists of: Constant region (Fc) with hinge region antibody recognition region (Fab) Hinge region is flexible allowing movement between two antibody-binding sites Papain digestion splits antibody into three fragments: 2 Fab, one Fc
IgA Abundant in secretions protecting mucosal surfaces. Predominant Ig in colostrum (providing passive immunity to newborn) Half-life = 6 days Activate alternate complement pathway Exist as dimer joined by J-chain.	***IgM*** Largest immunoglobulin composed of 5 Y-shaped monomers held by J polypeptide Cannot traverse blood vessels Ist Ab produced in response to Ag Half-life 5 days
IgD Minor Ig Surface receptor for antigen in B-lymphocyte Monomer Have high carbohydrate content of numerous oligosaccharide units Function not clear	***IgE*** Present in traces Monomer Found in spleen, tonsil, adenoids and mucosal membranes of lungs and GIT Elicit allergic response by inducing mast cells to release histamine.

	Provide immunity in mucosal membrane and parasites.
IgG Most abundant Ig (75% of total Ab (antibody in serum) Composed of single Y-shaped monomer: 2H2L Molecular mass 160kDa Half-life 22 days Present in all extracellular fluid	***Monoclonal Ig*** Produced by single B cell arising from B cells' malignant or benign transformations It reacts only with a single epitope It may be normal or fragmented or truncated It produces a single band on get electrophoresis It is associated with myeloma, Waldenstrom's macroglobulinemia and benign monoclonal gammapathies of uncertain significance (MGUS).
Multiple Myeloma Malignant proliferation of monoclonal plasma cells producing monoclonal antibodies with kappa or γ- light chains (Bence Jones protein) and identification of bone (skull, vertebrae, rib, pelvis) and organs. Bence Jones proteins excreted in urine and show M-band in γ-globin region on electrophoresis.	**Binding of Antibodies** Antibodies bind to specific molecules through their hypervariable loops. Variable domains in L and H chains (V_2 and V_H) come together at the ends of arms extending from the structure.
Antibody Biosynthesis Lymphocytes have the ability to make extremely large repertoire of antigen receptors and soluble antibodies. Wide range of immunoglobulins (Ig) are produced by genetic recombination and additional mutations	About 10^8 different antibody variants occur in every human being. Various Ig types that arise from combination of five different types of heavy (H) chain *(α, δ, $\in$, γ, μ)* and two types of light *(L) chain (K and y)* are known as *isotypes.*

during the development and maturation of individual lymphocytes.	During immunoglobulin biosynthesis, plasma cells can switch from one isotype to another (termed as gene switch).

Level of Response	***Outcome***	**Immunologic Dysfunction**
Decreased Response	Immunodeficiency: Primary Secondary	Immune system's activities are mostly beneficial; however, there are several situations where they can have deleterious effects. Disorders of immune regulation can be aberrations of quantity, quality or direction of response. Autoimmune diseases can arise from a malfunctioning immune system.
Increased response	Tuberculosis Leprosy Immune complex disease Lymphoproliferative disease (monoclonal response)	
In appropriate response	Allergic disease: IgE	
Increased and Inappropriate Response	Allergic disease Auto immune disease (mono clonal response)	

Antibody Deficiency Disorders:	Deficiency in antibody production results in agammaglobulinemia, hypogammaglobulinemia and specific immunoglobin deficiency.	
1. Infantile X-linked or Bruton's Agammaglobulinemia	X-linked recessive trait This appears in infants older than 5-6 months after infant's supply of IgG is depleted.	*Defect:* deficiency of all classes of antibodies due to failure of bone marrow, pre-B-cells to mature to circulating

	Antibodies especially IgM are not produced.	antibody-producing B cells.
2. Transient Hypogammaglobulinemia of Infancy	This condition also appears at 5-6 months of age.	*Defect:* delayed production of IgG between 5-6 months of age, when maternal IgG decreases. Cause of disorder is not known.
3. Selective Ig Deficiency	It is the most common immunodeficiency disorder where patient can be symptom free or presents with recurrent pulmonary infections and gastrointestinal problems.	*Defect:* failure of B cells to mature to plasma cells that produce IgG. All other classes of antibodies all normal in these patients.

Terminology Continued		
Antigens	Substances that can be identified by B cells' immunoglobulin receptor or by T cells' receptor when integrated with MHC.	
Antigens and Antigenic Receptors	The molecule that initiates the production of an antibody is an antigen. The molecular surface characteristic present of an antigen that can be bound to an antibody is an epitope.	
Antigens are Classified as	Endogenous antigens (generated within cells)	Exogenous antigens (entering from outside)
Antigenic specificity is the potential of host cells to identify an antigen by its exclusive molecular structure.		
Examples of Antigens	1. Microorganisms a. Differing epitopic sequences are the basis for their serologic classification	

	b. Vaccines frequently contain components responsible for virulence of the organism that are inactivated c. Streptococcal antigens are exemplified by group-specific carbohydrates, type-specific M proteins, streptolysin O, erythrogenic toxins, and multiple enzymes. 2 . Human tissue antigens are exemplified by blood group antigens, organ specific antigens, and leukocytic antigens (HLA)
Epitope (Antigenic Determinant, Ligand)	a. Defines a short sequence of sugars or amino acids in an antigen molecule which binds to a hypervariable region on the antibody b. Is usually repeated several times, and the number of repeats on the antigen is referred to as the valence A microorganism may have multiple, different epitopes
Hapten	a. A portion of the antigen molecule that contains the epitope b. Reacts specifically with its antibody but is incapable of inducing antibody synthesis without a carrier molecule.
Superantigen	a. Represented by certain retroviral proteins and bacterial toxins (*e.g.*, staphylococcal enterotoxins and toxic shock syndrome toxin 1) b. Can link multiple T cells to the MHC of APC via T-cell regions independent of their specific peptide binding sites c. Results in activation of many nonspecific T cells and APC, causing secretion of extraordinary amounts of cytokines.
Thymus Independent (TI) Antigens	a. Activate B cells without Th cell involvement b. Represented by multiple branched lipopolysaccharides (LPS) as are found on Gram-negative bacteria c. Activate B cells polyclonally without regard to B-cell specificity (B-cell mitogens)

<table>
<tr><td colspan="4">Vaccines</td></tr>
<tr><td colspan="2">Vaccination brings about protective response to counter the target pathogen via immune system without any chance of getting the disease and associated possible complications.

Innate immune response is initiated by vaccines, which then induces an antigen-specific response from adaptive immunity.</td><td colspan="2">Goal of vaccination is immunological memory and is accompanied with presence of antibodies and creation of memory cells which quickly reactivate on consequent exposure to the known pathogen.

To understand the mechanism of action of vaccines, it is essential to predict their safety profile, their efficacy, and their predicted benefit for the individuals vaccinated and the population in general.</td></tr>
<tr><td colspan="4">On the basis of the biology of the infection, and the targeted disease to be prevented, the vaccine may need the help from the different immune mechanisms to be effective, like humoral immunity (i.e. antibodies) or cell-mediated immunity (i.e. T cells) or adaptive immunity.</td></tr>
<tr><td colspan="4">Different Types of Vaccines</td></tr>
<tr><td>Different processes may be employed in production of Vaccines.

<u>Vaccines may Contain</u></td><td>Live attenuated viruses/ pathogens</td><td>Inactivated complete pathogens, toxoids (inactivated form of disease causing bacterial toxin)</td><td>Or parts/ regions from pathogens (like natural/ recombinant proteins/ virus-like proteins/ polysaccharides/ conjugated polysaccharides).</td></tr>
<tr><td><u>Types of Vaccines:</u></td><td>Non-adjuvant vaccine: containing only antigen as the key component.</td><td colspan="2">Adjuvant vaccine: having two key components - the antigen (pieces of virus or killed/ attenuated virus) and the adjuvant.</td></tr>
<tr><td>Concept of Vaccine Development</td><td colspan="2">Antigen – a chemical that triggers an immune response and causes antibody production in the body.</td><td>Adjuvant - a substance integrated to a vaccine, intended to enhance the immune reaction of the vaccine.</td></tr>
</table>

Vaccinations are created with live/ killed viruses, or molecular subunits from the virus.	A ***live vaccine*** contains a small dose of an active virus.	A ***killed vaccine*** is prepared with inactivated virus.

- It is probable, although infrequent, for live vaccines to lead to disease they're meant to prevent.
- Live vaccines may be prepared by culturing the virus *in vitro*, that may lead to mutations allowing them to better growth in the lab than *in vivo*, thereby impeding their potential to cause disease.
- Even if live vaccines are created to cause a few symptoms, there is possibility of back mutationscausing the virus to reinstate in the host and severe manifestation of the disease.

MUSCLE

Morphology of Muscle	Skeletal muscles are striated Consists of parallel bundles of fibers known as myofibrils.	Repeating unit called sarcomere constitutes each myofibril.
Sarcolemma: Plasma membrane of muscle fiber.	***Sarcoplasm:*** Cytoplasm in muscle fiber cell	***Sarcoplasmic Reticulum:*** ER in the muscle cell.
Sarcomere:	Consists of striated components between Z bands at the centre of I band. **EM** shows band structure of sarcomere: dark (A) bands and light (I) bands	
In A band lies a lighter zone H zone. In the middle of H zone is present a dark M line	A band contains thick filament I band contains thin filament These filaments are linked by cross bridges where they overlap.	Thin filament contains mainly actin Both thick and thin filaments do not change length or width during muscle contraction, but they overlap.

I band separates adjacent sarcomere along a dark, dense, marrow Z line. Thus, a sarcomere is Z line to – Z line repeat	Thick filament contains mainly myosin.	Then filaments are formed by end-to-end polymerization of globular actin subunit and referred to as F-actin, fibrous actin. Tropomyosin and troponin are associated with F-actin.

Actin

- 43,000 MW size, globular protein (globular or G actin).
- One molecule of ATP binds to each actin.
- Each G-actin has two binding sites, one for ATP and second to divalent metal Mg^{2+}.
- G-actin-ATP-Mg^{2+} complex aggregates to form F-actin.
- Exist as double stranded helix, their ends anchored to Z-disks of sarcomeres.
- Helices grooves of actin filament are occupied by tropomyosin.

Troponin (T_n)

- Complex of three polypeptides TnT, TnI, TnC.
- TnT binds to tropomyosin

Tropomyosin

- Rod shaped protein with two dissimilar subunits α and β.
- Forms aggregates in head to tail configuration
- Interact in flexible manner.
- One tropomyosin interacts with seven actin monomers
- One troponin complex bind to every tropomyosin

Myosin

- Thick filament contains myosin
- Myosin is long fibrous molecule with two globular heads on each end.
- 6 polypeptides: consists of 2 heavy chain and two pairs of different light chains.
- Light chains are calmodulin like proteins

<table>
<tr>
<td rowspan="3">
• TnI binds to actin, inhibits binding of actin to myosin

• TnC is calmodulin-like protein, binds Ca^{2+}.

• At high calcium concentrations found in contracting muscle (10^{-5}M), TnC undergoes a conformational change, releasing TnI from actin and causing tropomyosin to move deeper into its binding groove, exposing myosin-binding site of actin.

• There are 7 thin filaments surrounding each thick filament. Each thick filament interacts with multiple thin filaments.
</td>
<td colspan="2">• Two types: I, II</td>
</tr>
<tr>
<td>Type I not found in muscle cells</td>
<td>Type II required for muscle contraction</td>
</tr>
<tr>
<td colspan="2">
• N-terminal is globular structure to which L chains are bound

• Two H chains are attached through coiling of their α-helical tails in coiled tail structure.

• Tail sections are aligned in parallel manner on both sides of M line of H zone in sarcomere, with head group pointing towards Z line.

• N-terminal can bind to filamentous actin (F-actin)
</td>
</tr>
<tr>
<td colspan="3">
• Each thick filament contains 400 myosin

• Cross bridges between thick and thin filament are seen in A band by binding of N terminal head group of myosin to actin.

• Myosin head contains ATPase activity to provide energy for contraction and actin binding.

• In physiological conditions, numerous myosin molecules aggregate forming a thick filament having globular heads that project on sides of both the ends.
</td>
</tr>
</table>

Minor Proteins in Muscle	
α Actinin, desmin, vimentin	present in 2 disks to which thin filaments are anchored.
C protein, M protein	present in M disk, where thick filaments are anchored.
Titin	unusual protein: Longest known polypeptide chain

	3-6 molecules of titin associate with each thick filament between M and Z-disks Function as metabolic bunge cord to secure thick filaments positioned on sarcomere Compresses through muscle contraction; whereas during relaxation, resists sarcomere extension past the point where thin and thick filaments have started to slide past one other.
Nebulin	α helical protein associated with thin filament Act as template for actin polymerization
Dystrophin	Member of family of flexible rod-shaped proteins Functions to anchor specific transmembrane glycoproteins. In Duchenne muscular dystrophy, dystrophin gene contains deletions and individuals have no detectable dystrophin in their muscles.
Mechanisms of Muscle Contraction	
Sliding Filament Model	Each sarcomere of a myofibril shortens and thin filaments slide across the thick filaments. Shortening of sarcomeres result in significant reduction in length of myofibrils. Each myosin cross-bridge to actin repeatedly detaches and reattaches itself and free energy for this is provide by ATP hydrolysis.

Basic Contraction: Four Stages			
Myosin along with, ADP and Pi bound to it diffuses to make contact with actin and binding with actin	This binding induces conformational change in myosin Actin pulled across thick filament, ADP	ATP binds to myosin, releasing it from actin.	Hydrolysis of ATP by myosin and conformation change in myosin occurs so

becomes tight following loss of Pi.	released, conformation of myosin changes.		that it is free to bind actin.
This cycle continues till Ca^{2+} and ATP are present.			

Energy Reservoirs for Muscle Contraction	
1. Creatine Phosphokinase (CPK) Phosphocreatine + ADP $\rightleftharpoons$ ATP + creatine If metabolic activity is insufficient to keep up with the need for ATP, CPK helps to maintain constant cellular levels.	**2. Adenylate Kinase** Provides additional ATP *via*: 2ADP $\rightleftharpoons$ ATP + AMP ***Regulation*** Key regulatory molecule: Ca Calcium is sequestered in sarcoplasmic reticulum.
Presence of Ca Ca binds to TnC, causes conformation change in Tn-Tropomyosin complex. Tropomyosin moves away from myosin binding site on actin, allowing myosin to bind and carry out contraction.	***Absence of Calcium*** Tropomyosin lies in groove of actin helix and myosin ADP-P complex cannot bind

- On electrical excitation, depolarization of sarcolemma results in release of calcium from sarcoplasmic reticulum.
- Calcium is removed from sarcoplasm and tropomyosin-Tn complex conformation reversed
- Tropomyosin returns to groove on actin helix, inhibiting binding of myosin to it.

NERVE

<table>
<tr><td>Cells of the Nervous System:

Neurons</td><td colspan="3">Neurons are the information-processing cells of the nervous system. There are two broad categories of cells in the nervous system:
(a) Neurons: that process information
(b) Glia: that provide mechanical and metabolic support to neurons.</td></tr>
<tr><td>Three general categories of neurons:</td><td>Receptors

Highly specialized neurons that begin the process of sensation and perception. e.g., photoreceptors.</td><td>Interneurons

Serve to process information in many different ways, constitute the bulk of human nervous system</td><td>Effectors or Motor Neurons

Send signals to the muscles and glands of the body.</td></tr>
<tr><td>Structure

A characteristic neuron constitutes three distinct regions: a cell body, the dendrites, and an axon</td><td>Cell Body

Contains nucleus and associated intracellular structures.</td><td>Dendrites

Specialized extensions from the cell body that primarily function to transmit information from other cells/ neurons to cell body</td><td>Axon

It transmits the action potentials from the cell body/ soma to its terminal endfeet, that synapse with other neurons or effector cells/organs.</td></tr>
<tr><td colspan="4">Many neurons possess an axon, that transmits information from the soma to other neurons/ other cells, but many small neurons do not.

Axons terminate into endfeet, also known as terminal boutons (buttons), that transmit information to the receiving cell/ neuron through synapse.

Both dendrites and axons are extensions of the cell body, and are also known as processes.</td></tr>
<tr><td colspan="4">The site of Communication Between Two Neurons is termed - ‘Synapse’</td></tr>
<tr><td>Cell Membrane of Neuron</td><td colspan="3">Separates neurons from other types of cells and also from the extracellular fluid. It is critical in understanding the functions of neurons.</td></tr>
</table>

	Major structural constituents – phospholipids/ fatty acids and embedded protein molecules (integral proteins that perform selective transport of molecules across membrane like for glucose; peripheral proteins that perform many functions.	
Glia and other Supporting Cells	They perform a various housekeeping function for the brain. Glial cells hold the brain together, and occupy space between the neurons.	
Two Types of Glial Cells	Large-bodied macroglia: astrocytes and oligodendrocytes.	Smaller microglia: perform "housekeeping" functions, removal of dead cells within the brain
Astrocytes	***Oligodendrocytes***	***Schwann Cell***
Show characteristic absence of organelles within the cell bodies, and offer structural support for neurons in brain. Also aid in neuronal repairing after damage to brain	Cell bodies of these cells contain organelles, and many microtubules arranged in parallel arrays. They are involved in producing myelin, meant to surround axons in many neurons and plays critical roles in the CNS.	Also a supporting cell type and similar to the oligodendrocytes. They first encircle an axon, and then wrap itself around the neuron, creating a *myelin sheath* ***Myelination*** Greatly enhances the velocity with which an axon carries the action potentials.
Nerve Cell Action Potentials and Synaptic Transmission (Fig. 15.6)	Neurons connect with each another at junction sites called synapses.	At the synapse, the neuron directs a message to nearby target neuron or other cell type.
Types of synapses:	Chemical: Most of the synapses are typically chemical and communicate with help of chemical messengers.	Electrical: flow of ions is direct through in these type of synapses.

<table>
<tr><td>Action potential:</td><td>At the chemical synapse, presynaptic neuron is triggered by action potential to release neurotransmitters.</td><td>The molecules of neurotransmitters bind with receptors present on postsynaptic cell resulting on firing of an action potential.</td></tr>
<tr><td>Many synaptic vesicles are present inside a presynaptic neuron. These membrane-bound spheres are filled with molecules of neurotransmitter.

→</td><td>→A small gap lies between the presynaptic neuron axon terminal and postsynaptic cell membrane. This gap is termed as synaptic cleft.

→</td><td>→As the action potential, i.e. nerve impulse, reaches the axon terminal, it induces voltage-gated calcium channels of the cell membrane.

↓</td></tr>
<tr><td rowspan="2">→The molecules of neurotransmitter diffuse across the synaptic cleft and bind to receptor proteins present on the postsynaptic cell. Activation of the postsynaptic receptors leads to either opening or closing of ion channels of the cell membrane. This process may be depolarizing i.e. making the cytoplasmic side of membrane more positive — or hyperpolarizing i.e. making the inside of the cell more negative — all depending entirely on the ions involved.</td><td>→Ca^{2+} synaptic vesicles to fuse with the axon's endfeet/ terminal membrane, releasing neurotransmitter into the synaptic cleft.

←</td><td>→Ca^{2+} which is present at a much higher concentration outside the neuron than inside, rushes into the cell.

←</td></tr>
<tr><td colspan="2">Sometimes, these changes on channel behaviour may be direct, i.e. the receptor is a ligand-gated ion channel. Or else, the receptor may not be an ion channel itself rather activates other ion channels via signalling cascades.</td></tr>
<tr><td>Excitatory and inhibitory postsynaptic action potentials:</td><td colspan="2">When a neurotransmitter molecule binds to the postsynaptic cell membrane, it leads to opening or closing of the ion channels, leading to a localised alteration in the membrane potential i.e. the voltage across the cell membrane, of the post synaptic or receiving cell.</td></tr>
<tr><td colspan="2">Excitatory Postsynaptic Potential, or EPSP</td><td>Inhibitory Post-Synaptic Potential, Or IPSP</td></tr>
</table>

The altered localised action potential leads to spontaneous firing of the target cell's action potential.	The altered localised action potential leads to less likely firing of the target cell's action potential.
EPSPs are Depolarizing, making cytoplasmic side of cell membrane more positive, and bringing membrane potential nearer to the threshold for firing of an action potential. Occasionally, a single EPSP may not be large enough to bring the neuron to its threshold, but it may sum-up through other EPSPs triggering an action potential.	IPSPs have the opposite effect. The membrane potential of postsynaptic neuron is kept below threshold for firing of an action potential. IPSPs are also critical as they can cancel out or counteract, the excitatory effects of EPSPs.

Postsynaptic Receptors	Neurotransmitter molecules of the synaptic cleft diffuse through the gap region to the post-synaptic cell's membrane, where they may bind with two kinds of post-synaptic receptors.	
Name of the receptor	Inotropic receptors	Metabotropic receptors
Type of the receptor	Ligand gated ion channels	G protein coupled receptors
Response of the receptor	The ion channel permits ion flux to alter cell voltage	Receptor acts via secondary messengers causing cellular effects
Speed of the response	Rapid/ quick	Slow
Length of the response	Short-acting	Prolonged response
This may lead to either depolarisation *i.e.* promoting or hyperpolarisation *i.e.* inhibiting generation of action potential in a post-synaptic neuron.		

Inactivation/ Removal of Neurotransmitters	After the post-synaptic membrane responds, the neurotransmitter molecule in the synaptic cleft may either be inactivated or removed, in following ways:
Re-uptake	*E.g.* serotonin is packaged back into the pre-synaptic neuron via transporter membrane proteins. The neurotransmitter molecules may

	either be recycled through re-packaging into vesicles or may be broken down by enzymes.
Breakdown	*E.g.* acetylcholine is inactivated by enzymatic break down by acetylcholinesterase present in the synaptic cleft
Diffusion into the surrounding areas	
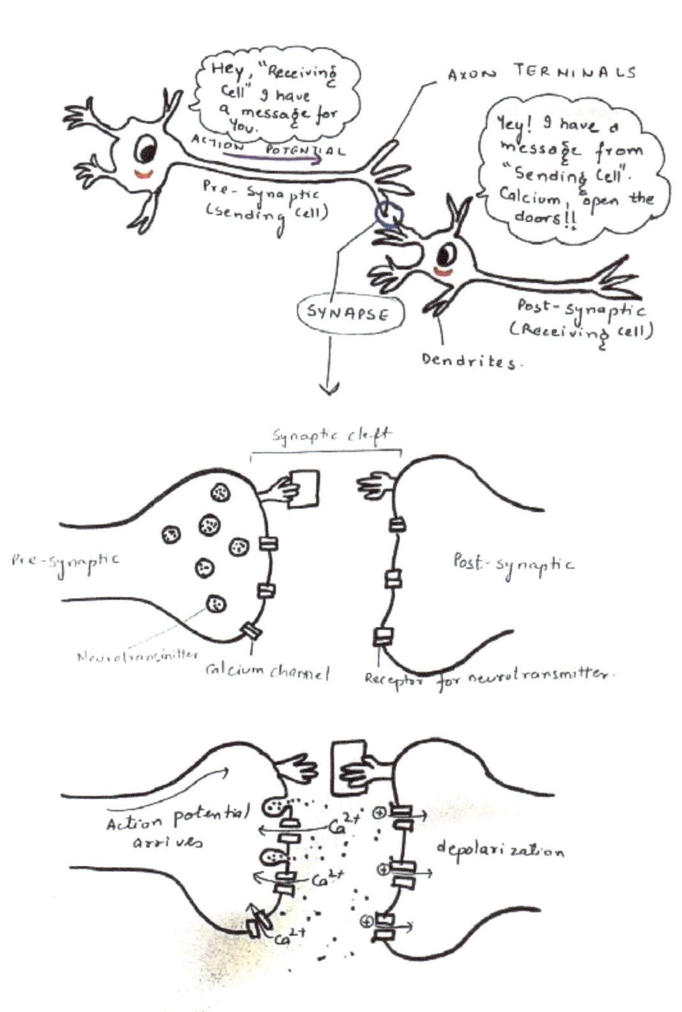 **Fig. (15.6).** Action potential in a Nerve Cell.	

EYE

<table>
<tr><td colspan="4">Tear Film (Fig. 15.7)</td></tr>
<tr><td colspan="2">Functions</td><td colspan="2">Structure</td></tr>
<tr><td colspan="2">• Keeps moist the cornea and conjunctiva serving as a lubricant for lids
• Transfers air/O_2 to the cornea
• Prevents infections
• Washes away irritants and debris
• Provides pathway for WBC in event of injury</td><td>1. Mucin Layer
• 0.02-0.05 µm, deepest layer
• Secreted from conjunctival goblet cells, of main lacrimal gland
• Made of glycoprotein, mixed with lipids, anchored by microvilli</td><td>• Balances tear film, Semisolid sand highly hydrated.
• Lubricates the ocular and palpebral surfaces, reducing friction.
• Slippery film over foreign particles, protecting cornea and conjunctiva to counter abrasive effects.</td></tr>
<tr><td>2. Aqueous Layer</td><td colspan="2">• 10 µm, forms a uniform layer : middle layer, Secreted from lacrinal and accessary glands of Krause & Wolfring
• Lysozyme, lactoferrin, tear specific prealbumin and immunoglobulin A</td><td>• Bicarbonate and proteins give buffering capacity to tears
• Macromolecular mucoglycoproteins determine the surface tension of fluid
• Delivers atmospheric O_2 to the epithelium.
• Washes away irritants and debris.</td></tr>
<tr><td>Chemical Composition</td><td colspan="2">Water
• Major component, 98.2%
• With salts dissolved: K^+, Na^+, HCO_3^-, Cl^-, Ca^{2+}</td><td>Proteins
• Un-stimulated tears: 2gm/100ml
• Stimulated tears: 0.3-0.7 gm/100ml</td></tr>
</table>

<table>
<tr><td></td><td><u>Albumin</u>

* Tear specific protein (prealbumin)

• Exact function is not known, acidic protein, might stabilize thin tear film i.e. Tear Film C</td><td><u>Immunoglobulins</u>

• IgA is most prominent

• Secreted by plasma cells of conjunctiva

• Defend against bacterial or viral antigens.

• IgE and IgM are also seen.</td></tr>
<tr><td></td><td><u>Lysozyme</u>

• Glycosidase enzyme, with net positive charge

• Produced by Lacrinal gland's acinar cells

• Work to counter bacterial infections and bacterial cell wall lysis</td><td><u>Glycolytic Enzymes, Lactate Dehydrogenase</u>

Betalysin: antibacterial agent

Mucopolysaccharides

Glycoproteins

Amino acids

Lipids</td></tr>
<tr><td></td><td><u>Metabolites</u>

Lactate, glucose, pyruvate, urea</td><td><u>Electrolytes</u>

K^+, Na^+, Ca^{2+}</td></tr>
</table>

<table>
<tr><td colspan="3">Aqueous Humour</td></tr>
<tr><td>Functions

• Maintenance of intraocular pressure

• They also have metabolic role:</td><td>1. Cornea – takes glucose and oxygen, release lactic acid and carbon dioxide

2. Lens – use oxygen, glucose, amino acids and potassium and release lactate, pyruvate and sodium</td><td>• Optical Function

Act as a diverging lens of low power

• Clearing functions</td></tr>
</table>

Biochemical composition:	* Water 99.9%	Proteins - colloid content, 1. 5- 16 mg/100ml (plasma 6-7gm/100ml) because of blood aqueous barrier
	IgM and IgG but NO IgD and IgA	
	Free amino acids	Growth factors like FGF, TGF beta, IGF-1
Non colloidal contents – dissolved solids:	1. High vitamin C 2. High Pyruvate and lactate	3. Na, K, Ca, Mg, Cl, HCO3 4. Low glucose and urea compared to plasma
Biochemical composition (Fig. **15.8**):	Anterior chamber: • Lesser bicarbonate • Higher chloride • Lesser ascorbate • pH 7.6 The difference is due to diffusional exchange across the iris	Posterior Chamber: • Higher bicarbonate • Lower chloride • Higher ascorbate • pH 7.57 Iris vessels are permeable to anions and non-electrolytes

Aqueous Humour: Biochemistry of Formation		
Active Transport of		***Passive Transport Across NPE***
• Sodium • Chloride • Potassium	• Ascorbic acid • Amino acids • Bicarbonate	• Water • Chloride • Sodium

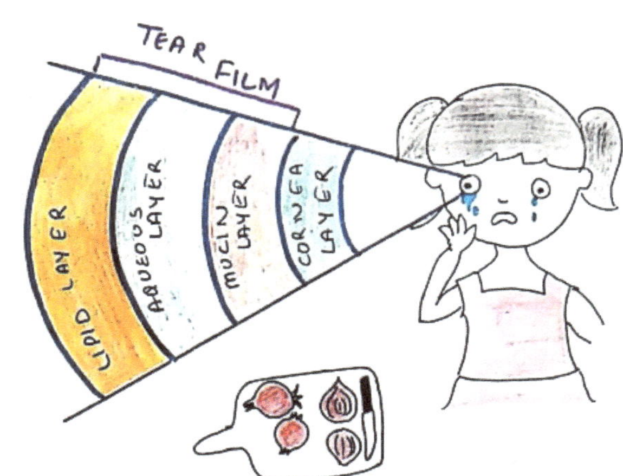

Fig. (15.7). Composition of a tear film.

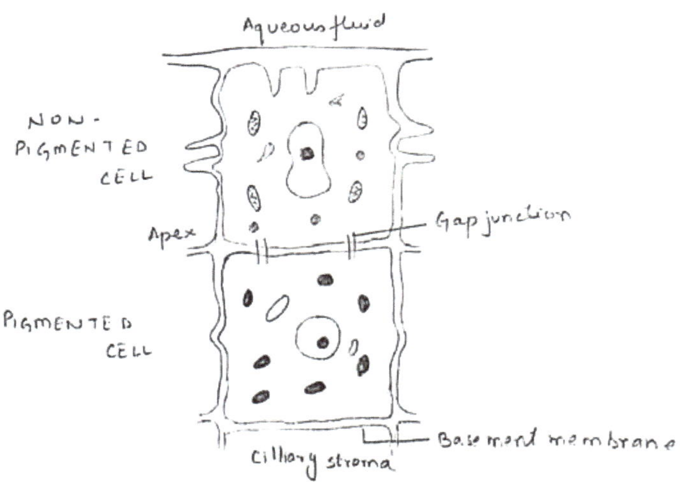

Fig. (15.8). Ciliary Epithelium and Aqueous fluid.

Cornea:	Most peripheral region of the eye and the key physical barrier towards external factors having higher	Important nutrients and oxygen reach the cornea from the limbal blood vessels, tear film, and aqueous humor; the aqueous humor is the main source of nutritiion. Despite the absence of normal blood vessels source, the biochemistry of cornea is non - neutral to the changes in

	metabolic activity in the endothelium and epithelium.	metabolism associated with chronic systemic diseases or leading to death.
Stroma	• Collagen fibrils 70% • Soluble proteins (Ig G, A, M) • Proteoglycans: Keratin sulphate,	• Enzymes • Matrix metalloproteinases (MMP -1, -2 & -3) • Electrolytes
Epithelium	*. Water 70% • Electrolytes (Na^+, K^+, Cl^-)	• Enzymes necessary for metabolism • Acetylcholine
Lens	Lens water – Dehydrated: Lens contains less water compared to any other tissue • Amino acids • Carbohydrates • Lipids • Electrolytes • Glutathione	Soluble Crystallins and insoluble albumiboids: Insoluble and soluble proteins incessantly upsurge in total amounts Metabolically active crystallins are enhance in direct proportionally with increase in the lens surface Metabolically inactive albumoid is accumulated proportionally with square of weight of lens

Vitamin A (Retinol) (Fig. 15.9)

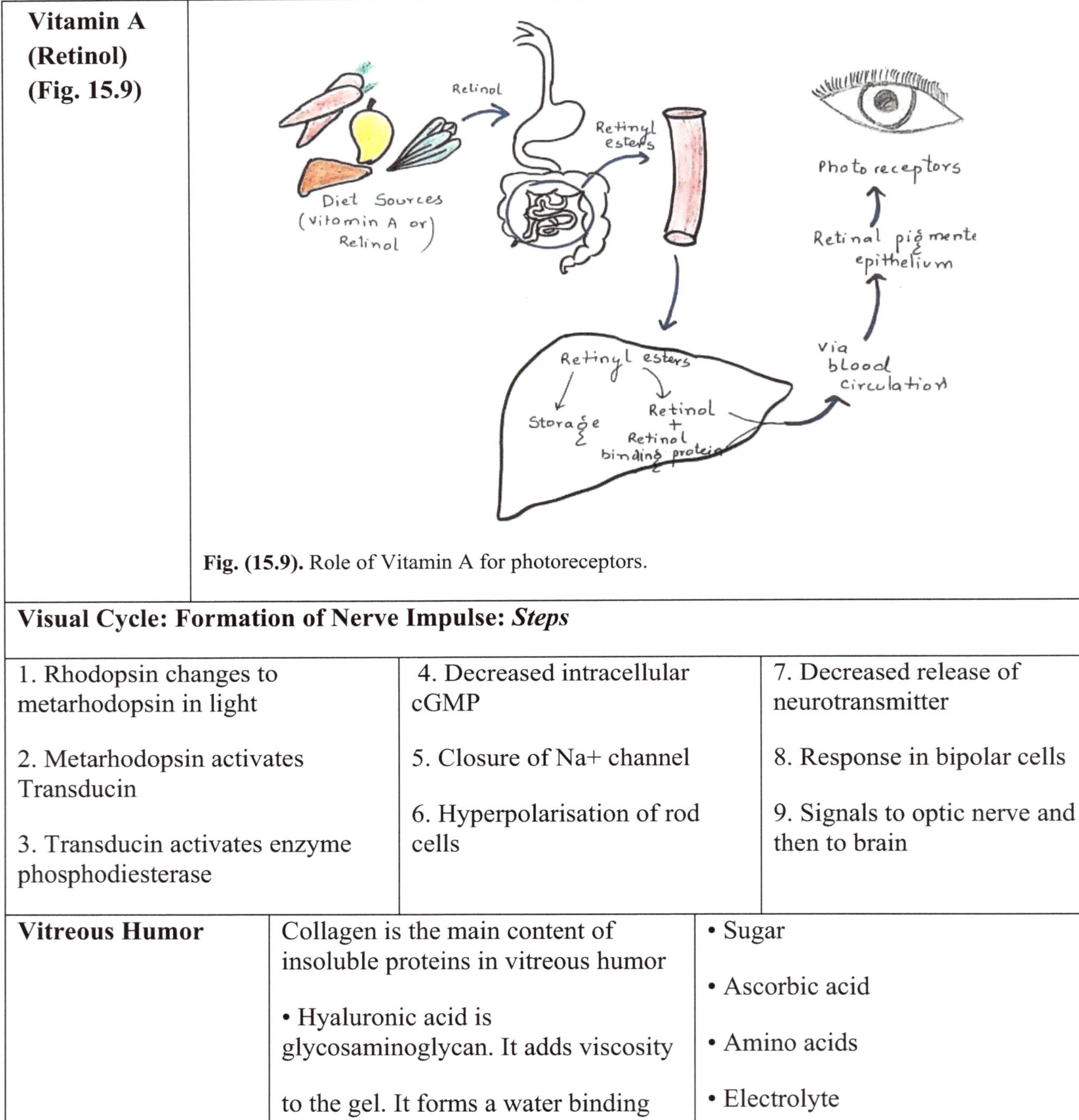

Fig. (15.9). Role of Vitamin A for photoreceptors.

Visual Cycle: Formation of Nerve Impulse: ***Steps***		
1. Rhodopsin changes to metarhodopsin in light 2. Metarhodopsin activates Transducin 3. Transducin activates enzyme phosphodiesterase	4. Decreased intracellular cGMP 5. Closure of Na+ channel 6. Hyperpolarisation of rod cells	7. Decreased release of neurotransmitter 8. Response in bipolar cells 9. Signals to optic nerve and then to brain

Vitreous Humor	Collagen is the main content of insoluble proteins in vitreous humor • Hyaluronic acid is glycosaminoglycan. It adds viscosity to the gel. It forms a water binding meshwork around collagen fibers	• Sugar • Ascorbic acid • Amino acids • Electrolyte

<table>
<tr><td></td><td colspan="2">• Soluble proteins has mainly glycoprotein and albumin</td><td></td></tr>
<tr><td colspan="4">Human eye has the potential of detecting an assortment of colours, creating images of things miles away, and reacting to as less as one single photon of light.

Rods and cones are not present in the optic disc, and this portion present on the lower periphery of the retina is a *blind spot*, *i.e.* no light is detected focused on that part.

Fovea or "yellow spot" is the site of most acute vision, and contains few rods and many cones.</td></tr>
<tr><td>**Aqueous Humour:**</td><td colspan="3">Ciliary body constantly produces the clear, watery aqueous humour that fills the front cavity of the eye.</td></tr>
<tr><td colspan="4">The back cavity, containing jelly-like vitreous humour makes for most of the volume in the eye.</td></tr>
<tr><td colspan="2">Vitreous and Aqueous humour work as liquid lines which help to focus the light onto retina.</td><td colspan="2">Lens is in itself a transparent type of protein disc which focuses the image on retina</td></tr>
<tr><td>Refractive index:</td><td>Cornea: 1.38</td><td>Lens: 1.42</td><td>Vitreous and Aqueous humours: 1.33.</td></tr>
<tr><td colspan="4">Main difference in the refractive index happens between air and the cornea, which is essential for formation of image.

The accurate and delicate control is attained by lens that functions for fine adjustment.</td></tr>
<tr><td>*Photoreceptors: Rods and Cones*</td><td colspan="3">Millions of photoreceptor cells are present in retina called as rods and cones.

125 million rods + 6 million cones (approximately) are found in Human retina.</td></tr>
<tr><td colspan="4">Rod cells are profusely found towards the retinal periphery while cone cells are largely found in the central region of retina

Each rod or cone cell contains an outer segment having a stack of folded membranes/ discs, that contain embedded visual pigments.</td></tr>
</table>

Visual pigment present in rods is processed into membranes of flattened vesicles found in outer segment, called as *rhodopsin.* Rod cells show higher sensitivity to minor exposures to light. Rods possess their own form of opsin, that combines with the conjugated polyenal (*11-cis-retinal)* present in the retina forming red-purple (11-cis-imine), also called *rhodopsin or "visual purple"* *Rhodopsin* consists of 348 amino acid residues organized as seven alpha helical hydrophobic segments that pass amongst two sides of the membrane of photoreceptor. The rhodopsin irradiation leads to a number of succeeding conformational changes observed as appearing and disappearing sequences of intermediates of variable colours: *rhodopsin "bleaching"* Signal transduction pathway from reception of light in a rod cell to creating receptor potential.	In cones, visual pigment is known as iodopsin. Three categories of cone cells: possess different types of iodopsin and respond to lights of distinct wavelengths: One responds ideal to red light One responds to green The other responds to blue. Generally, colour-detection results from integrative degree of excitation from three kinds of cones. The white light sensation is observed only when all the three kinds of cones are excited equally. Cone cells provide sharper images.
Blood Supply Function: optimal way of nutrients' delivery and removal of waste products. *Dysfunction of blood supply:* Causes spontaneous dysfunction or damage occurs to eye.	*E.g., Chronic systemic diseases*: May lead to upsurge in toxic metabolites in blood plasma: may also affect and harm retina *Diabetes*: High glucose levels cause changes in metabolism that influences the functions of the retina in long-term period *Dietary mistak*es: Accidental consumption of methanol. May cause irreversible damage.

Retina	*Unique:* Highest consumption of oxygen per unit wt. of any tissue in the human body Has two distinct circulatory systems that meet the metabolic demand	*Biochemical differences exist between:* Vascular (iris, ciliary body, and retina) and a-vascular tissues of eye (lens and cornea) Most characteristic differences exist in cornea and retina.
Cornea	Key physical barrier towards external factors, and most outer part of the eye and the Has increased metabolic activity in endothelium and epithelium: Has high levels of ATP compared to other eye tissues, requires additional energy reserves in cells.	

Role of High ATP Content

Protection of eye against environmenttal factors.

Maintains integrity of corneal tissue.

Lack Regular Blood Vessels Supply

Important nutrients & Oxygen reach cornea from limbal blood vessels, tear film, and aqueous humor

Aqueous humor is main nutritive source

Lens

Avascular

Dependent only on aqueous humor as key nutritive source

Antioxidant Glutathione is produced primarily in the lens epithelium

High glutathione levels in lens minimizes the effects of free radical species predominant in aqueous humor (or produced by UV) on lens

Ciliary Body and Iris

Have different functional and anatomical structures

Identical major arterial circle situated in ciliary body supplies both the structures

Source of Nutrients in the Eye	Clear fluid derived from the blood plasma

Aqueous Humor	Free amino acid levels nearly equivalent to blood plasma Protein content of aqueous humor is less than 1% of that found in the plasma.
Applied Aspects	
Disruption in Bloodstream	Can be caused by presence of potential antigens in blood These accumulated proteins cause turbidity→ scatters light→ degrade optical efficiency of eye
Cornea	Biochemical content of cornea in liver cirrhosis and cardiovascular disease differ significantly as compared to healthy cornea Shows metabolic changes related to chronic systemic diseases and free radical species despite being lack of regular blood vessels to cornea
Lens	Usual balance between imbibition and osmosis is upset in cataracts and autolyzed lenses Autolysis - dependent cataractous change Numerous deviations in biochemical parameters observed due to processes conditioning, resultant from autolysis, or involving, whose primary requirement is having local acidosis.

ORGAN FUNCTION TESTS

Liver

Functions of Liver	Liver is the 'metabolic factory' of the body. It has essential metabolic, synthetic and excretory functions.		
Normal Liver Function	***Metabolic*** Carbohydrate metabolism	***Synthetic*** ***Protein Synthesis***	***Excretory*** Detoxification of:

	Lipid and lipoprotein metabolism Protein metabolism	Most plasma proteins: (except Ig, RhF, complement) Blood coagulation factors: II, V, VII, IX, XI Primary bile acids Lipoprotein Urea Bilirubin glucuronide	Xenobiotics Drugs Toxic metabolic wastes (ammonia, bilirubin)

Liver Function Tests

Metabolite	↑/↓	*Enzyme Affected*	*Cause*
Ammonia	↑	Ornithine transcarbamoylase	Congenital deficiency Cirrhosis Reye's syndrome Hepatic failure
Lipids:			
Cholesterol	↓	Decreased hepatic synthesis	Cirrhosis
Proteins:			
Plasma proteins	↓	Decreased synthesis of protein	Cirrhosis chronic liver disease
Albumin	↓	Decreased synthesis of protein	Cirrhosis

lg	↓	Decreased synthesis of protein	Hepatic failure
Others:			
α 1 AT (alpha 1 antitrypsin)	↓	Decreased synthesis of protein	Congenital
Ceruloplasmin	↓	Deficient hepatic copper binding apoprotein	Wilson's disease
Clotting factors	↓	Decreased synthesis	Chronic liver disease
Carbohydrate:			
Glucose	↓	Decreased glycogen content	Cirrhosis acute liver failure
Bile salts	↑	Decreased secondary bile salts formation	Cholestasis

Jaundice	Jaundice is yellow discoloration of tissues due to bilirubin deposition	
Pre-hepatic	***Hepatic***	***Post-hepatic***
Hemolysis Ineffective erythropoiesis	Drugs Prematurity Hepatitis Inborn errors of bilirubin metabolism Cirrhosis	CA pancreas, Biliary tree Gallstone Biliary stricture

	Tumors	

Table: Lab Findings in Jaundice

	Pre hepatic	*Hepatic*	*Post hepatic*
Serum bilirubin			
Unconjugated	↑↑	↑↑	↑
Conjugated	-/↓/n	↑↑	↑↑↑
Urine bilirubin	-	-	+
Fecal urobilinogen	↑↑	↑	-
Fecal stercobilinogen	-	↓	-
Amino transferase	↑	↑↑	↑
ALP	N	n / ↑	↑↑
Albumin	N	n / ↑	N
Prothrombin time	N	n / ↑	n / ↑

Gastric Function Test In disease of stomach and duodenum, alterations of gastric secretions often occur. Gastric function tests may be of value in:	Diagnosis of gastric ulcer Exclusion of diagnosis of pernicious anaemia.	Presumptive diagnosis of Zollinger Ellison syndrome Determination of effectiveness of surgical vagotomy.

Commonly used tests		**Terminology**
1. Fractional test meal	-	
	Alcohol	

2. Examination of gastric contents after stimulation:	Caffeine	*BAO* – Basal acid output meq/L is based on 1 hour collection of gastric secretion, normal value: 1.3-4 meq/L *MAO* – maximal acid output meq/L serum of meq/L of acid secreted in one hour following injection of histamine/ pentagastrin normal value: 4.9-38.9 meq/L (10 times the BAO)
	Histamine (and augmented)	
	Insulin	
	Pentagastrin	
3. Tubeless gastric analysis	-	*Applied:* *Hyperacidity:* Free acidity > 45mg/L, occurs in duodenal ulcer, 50% of gastric ulcer *Hypoacidity:* seen in pernicious anaemia *Achlorhydria*: seen in carcinoma stomach, chronic gastritis and shows no response on histamine stimulation.

Kidney (Renal) Function Tests

Function of Kidney	*Excretion*: excretion of Urea Uric acid Creatinine Drugs	*Retention of:* Glucose Amino acid Protein
	Drug metabolism	*Production of hormones*: erythropoietin, rennin, 1, 25 DHD$_3$.

GLOMERULAR FILTRATION RATE (GFR)	*GFR depends on:*

Normal GFR is 120 ml/min approximately and is equivalent to volume of 170 L/24 h.	• Rate of blood flow • Balance between hydrostatic and oncotic forces in afferent arterioles and glomerular filter (tubular fluid).

Table Renal function test

Test	***Utility***	
Blood urea	Apart from renal function urea is dependent on factors such as protein intake rate of tissue break down	
Serum creatinine	Marker of decrease in (glomerular) renal function	
Clearance (Cl) Insulin Creatinine Urea	Exogenous Endogenous	Creatinine Cl is a prognostic indicator of renal function and is nearly equal to GFR
Cystatin C	Provide accurate assessment of moderate decrease in GFR.	
Tubular function: Urine concentration or dilution test Urine acidification	Tubular function Tubular function	

Urine examination

Routine:	Volume pH	Specific gravity Osmolality	Help to asses kidney function and diagnose disease

<table>
<tr><td colspan="2">Abnormal constituents:</td><td></td></tr>
<tr><td>Protein
Blood
Sugar</td><td>Ketone Bodies
Bilirubin</td><td>Present in urine in: Renal diseases (diabetes mellitus, jaundice, hereditary defects of amino acid transport)</td></tr>
</table>

<table>
<tr><td>Clearance</td><td colspan="2">Properties of Ideal Substance for Clearance</td></tr>
<tr><td>It is ml of plasma cleared off a particular substance by both the kidneys per minute.</td><td>1. It should be neither secreted nor reabsorbed by the kidney.
2. It should not be affected by rate of urine flow.</td><td>3. Its levels in blood should remain constant.
4. Its clearance should be equivalent to GFR</td></tr>
</table>

<table>
<tr><td colspan="2">Table Comparison of Urea and Creatinine Clearance</td></tr>
<tr><td>Urea clearance</td><td>Creatinine clearance</td></tr>
<tr><td>1. Its value is 3/5th of GFR</td><td>1. Values are nearly equal to GFR</td></tr>
<tr><td>2. Plasma values of urea are not constant</td><td>2. Its plasma values remain constant throughout the day over 24 hrs</td></tr>
<tr><td>3. Urea is both secreted and reabsorbed by kidney</td><td>3. Creatinine is neither secreted nor reabsorbed by kidney</td></tr>
<tr><td>4. It is affected by rate of urine flow</td><td>4. It does not depend on rate of urine flow</td></tr>
</table>

<table>
<tr><td colspan="4">Table: Tubular Function Tests</td></tr>
<tr><td colspan="2">Proximal</td><td colspan="2">Distal</td></tr>
<tr><td>a. Plasma:</td><td>b. Urine:</td><td>a. Plasma:</td><td>b. Urine:</td></tr>
</table>

Urea	Volume	pH	Volume
Creatinine	Specific gravity	Bicarbonate	pH
Uric acid	pH	Chloride	Sodium
Bicarbonate	Glucose	Potassium	Osmolality
pH	Amino acids		
Potassium	Uric acid		
Magnesium	Retinol binding protein		
Phosphate	α1 macroglobulin		

Urine Osmolality	Normal values are 300 to 900 mOsm/kg in case of average fluid intake.	
Normal:	> 850 following 12 hrs fluid restriction, implying normal concentrating ability of kidney. Does not exceed 300 mOsm/kg in complete ADH deficiency.	
Defects of Urinary Concentration	Diabetes insipidus CRF Hypercalcemia	Drug toxicity, *e.g.* Lithium Renal disease

Osmolality	Plasma osmolality is 285-295 mosmol/kg and sodium make largest contribution.
Osmolality = $2(Na + K) + \frac{urea}{2.8} + \frac{glucose}{18}$ (Osmol/kg)	

Regulation of Water and Electrolyte Balance:	Kidneys play a predominant role via hormones namely, aldosterone, ADH, rennin-angiotensin.
Hormone	***Action***
Aldosterone	Increase Na, decrease K and H^+ reabsorption
ADH	Increase H_2O reabsorption
Renin-angiotensin	Stimulate aldosterone secretion, constrict vascular smooth muscle

GENETIC DISORDERS

Genetic disorders are conditions caused in whole or in part by a change in the DNA sequence.		
Four Categories of Inherited Genetic Disorders	**Inherited Single Gene Disorders**	
Single gene inheritance Chromosome abnormalities Multifactorial inheritance Mitochondrial inheritance	Show different characteristics in genetic inheritance autosomal dominant X-linked inheritance autosomal recessive Examples: cystic fibrosis.	sickle cell anemia or sickle cell disease, alpha- and beta-thalassemias, fragile X syndrome, Marfan syndrome, Hemochromatosis, Huntington's disease.

Multifactorial Genetic Inheritance Disorders	**Abnormalities in Chromosomes**	**Mitochondrial Genetic Inheritance Disorders**
Alzheimer's disease, heart disease, arthritis, high blood pressure, cancer, diabetes, and obesity.	Turner syndrome (45, X0), Klinefelter syndrome (47, XXY), "Cry of the cat" or Cri du chat syndrome (46, XY or XX, 5p-).	Hereditary Leber's optic atrophy (LHON), a disease of the eye Myoclonic epilepsy having ragged red fibers (MERRF) Mitochondrial encephalopathy, stroke-like episodes (MELAS), lactic acidosis, and a erratic type of dementia.

Disorder	***Inheritance***	***Defect***	***Features***
Achondroplasia	Autosomal dominant	FGR3 or Fibroblast growth factor receptor - 3	Short limbs, low nasal bridge prominent forehead, legs and arms show redundant folds of skin
Alkaptonuria	Autosomal Recessive	Homogentisic acid oxidase	HOMOGENTISIC ACID accumulates
Cystic Fibrosis	Autosomal Recessive	Cystic fibrosis transmembrane regulator (CFTR) : compromised role of Cl^- channel	Pancreatic insufficiency, obstruction airway disease of lungs caused by thick mucus, incidence of lung infections
Diabetes Insipidus	X-linked recessive	*Central diabetes insipidus*: damage to hypothalamus or pituitary gland disrupts production or	Extreme thirst, polyuria, Dehydration Unexplained weakness

		release of vasopressin *Nephrogenic diabetes insipidus:* kidneys don't respond to vasopressin	Muscle pains Irritability
Duchenne Muscular Dystrophy	X-linked recessive	Dystrophin (DMD) - deletions	Gradual atrophy of skeletal muscle, impaired heart and respiratory musculature
Ehlers-Danlos syndrome	Autosomal Recessive	Defect in collagen synthesis or structure	Type I and III collagen affected: skin, ligaments and joints affected Hyperextensible, Hypermobile joints, predisposition to dislocation fragile skin
Hypercholesterolemia	Autosomal dominant	LDL receptor defect: Impaired LDL uptake	elevated LDL levels, beside increased levels of cholesterol, and incidence of cardiovascular disease and stroke
Fragile X Syndrome	X-linked dominant (females less severely affected)	5' untranslated region of FMR1 gene has *trinucleotide repeat expansion* (CGG) in (expansion exclusively occurs in mother)	Boys bear long faces, large ears, prominent jaws, and may also show mental retardation.

Gaucher's Disease	Autosomal recessive	B-Glucosidase	Lysosomal storage disease: hepatomegaly, splenomegaly, infiltration of bone marrow.
Glucose 6-phosphate dehydrogenase (G6PDH) deficiency	Recessive, X-linked (generally found among persons of Mediterranean and African ancestry)	Glucose 6-phosphate dehydrogenase	Anemia: caused by increased hemolysis Anemia brought about by sulfonamide antibiotics, oxidizing drugs, sulfones (like dapsone), fava beans, etc.
Hemochromatosis	Autosomal recessive	Chromosome 6 short arm has unknown gene	Enhanced dietary iron absorption with hemosiderin being accumulated in visceral organs. Cardiomyopathy, cirrhosis, diabetes, arthritis, and skin pigmentation.
Hemophilia A and B	X-linked	Affects males only *Type A*: It is the most common type & caused due to lack of factor VIII, *Type B hemophilia*: It is caused due to lack of factor IX	Bleeding disorder: blood cannot clot Extensive bleeding caused from a wound or cut. Spontaneous bleeding internally may also occur, particularly in muscles and joints.
Huntington Disease	Autosomal dominant	Huntingtin – CAG repeat	Typically shows progressive abnormalities in motor, psychiatric and cognitive

		expansion occurs within exon 1	functions. Chorea: involuntary non-repetitive jerks
Klinefelter Syndrome	50% of the cases are due to errors of paternal meiosis	47, XXY males	Males are sterile having long limbs and small genitalia, besides breast development, and contours of a feminine body; disabilities in learning
Lesch-Nyhan Syndrome	X-link	deficiency of HGPRT	Self-mutilation, usually fatal by 2 years' of age
Marfan Syndrome	Autosomal dominant (negative dominant effect)	microfibril-forming protein (found in connective tissue) is encoded by Fibrillin-1 gene (FBN1)	scoliosis, tall stature, heart (prolapse of mitral valve, dilatation of aorta, ascending aorta dissection, excessive elasticity, and hypermobility of joints.
Myoclonic Epilepsy with Ragged Red Fibers (MERRF)	Maternal transmission, heteroplasty	Mutation of mitochondrial DNA in tRNA-lys gene	Dementia, myopathy, myoclonic seizures, ataxia, and deafness
Myotonic Dystrophy	Autosomal dominant	Untranslated 3' region of the protein kinase gene (DMPK) gene has CTG repeat expansion	Cardiac arrhythmias, muscle weakness, cataracts and atrophy of testicles in males. typical open triangle-shape of mouth
Neurofibromatosis I	Autosomal dominant	Microdeletion, involving NF1 gene, at 17q11.2	Numerous benign tumors of peripheral nervous system like neurofibromas, increased occurrence of malignancy (astrocytoma, neurofibrosarcoma, childhood

			chronic myelogenous leukemia and Schwann cell cancers)
Osteogenesis Imperfecta	Usually autosomal dominant	Genes encoding α1 chain or α2 chain belonging to type I collagen	Associated with under mineralized and deformed bones that are prone to recurrent fractures.
Phenylketonuria	Autosomal recessive	Typically due to mutation in Phenylalanine hydroxylase gene (PAH)	Mental retardation, if untreated
Polycystic Kidney Disease	Autosomal dominant	Mutations in polycystin-2 (PKD2) gene or polycystin-1 (PKD1) gene	Heterozygous individuals are predisposed Blood in urine, multiple renal cysts, kidney failure, and end-stage renal disease.
Prader Willi/Angelman (PWS/AS)	Multifaceted effects due to genomic imprinting from parent of origin	PWS region deletion and AS gene deletion located at 15q11-q13.	characterized by short stature, severe mental retardation, spasticity, seizures, and a characteristic stance, obesity, excessive and indiscriminate gorging, small hands, feet, hypogonadism and mental retardation.
Tay-Sachs Disease	Autosomal recessive	B-Hexosaminidase (Isoenzyme, HEXA)	spasticity, Hypotonia, blindness, seizures, death ~ age of 2. A retinal cherry red spot is an early indication.
Thalassemias	Autosomal Recessive	Point mutation leading to	Severe anemia

		premature chain termination ◦ decreased synthesis α or β chains of hemoglobin	
Turner Syndrome	Typically from error in transmission of paternally inherited sex chromosome	45, X females	Webbed neck, short stature, sterility, and broad chest having broadly spaced nipples. Normal IQ.
Wilson disease	Ceruloplasmin deficiency, copper built up	X chromosome) recessive	Copper builds up in the liver, brain, eyes and other organs. The extra copper can lead to organ damage that may cause death
Xeroderma pigmentosum	Autosomal recessive	One of the 9 genes of nucleotide excision repair (locus heterogeneity)	premature skin aging, acute photosensitivity, premalignant actinic keratoses, and malignant and benign skin neoplasms, like basal and/or squamous cell carcinoma, melanoma is developed in 5% of patients.
Diagnosis	Cytogenetic analysis: Karyotyping		Molecular analysis: Direct diagnosis of gene Indirect diagnosis of DNA *i.e.* linkage analysis

QUESTIONS

1. Explain the mechanism of muscle contraction.
2. What is sliding filament theory?
3. Describe Calcium as regulatory molecule.
4. Name the different muscle proteins.
5. Explain why fetal Hb has higher oxygen affinity than adult Hb.
6. Name Heme containing proteins.
7. Give the importance of glycated Hb.
8. Write a short note on LFT & RFT.
9. What are causes of jaundice and how will you differentiate them with biochemical investigations?
10. Why sickle cell trait individuals are protected against falciparum malaria?
11. What are different types of porphyrias?
12. Discuss chemistry and function of Hb.
13. Short notes:

a. Abnormal haemoglobin	b. Thalassemia	c. Sickle cell anaemia.
d. Serum Albumin	e. AG ratio	f. Nephrotic syndrome
g. Multiple myeloma.	h. Blood buffers	i. Acidosis
j. Alkalosis	k. Icterus	l. Jaundice

m. Obstructive jaundice
n. Kernicterus.
o. Blood in urine
p. Protein urea
q. Ketonuria
r. Bile pigments
s. Urea clearance
t. GFR
u. Tubular function test.

14. Discuss synthesis and degradation of Hb.
15. Discuss various plasma proteins
16. What is normal pH of blood discuss regulation of normal pH.
17. Give an account of normal constituents of urine and abnormal pathological constituents that may be present in urine
18. What is Antibody diversity? Enumerate some immune deficiency disorders

BIBLIOGRAPHY

Denise R Ferrier. Lippincott illustrated reviews: biochemistry. 7th Edition. Philadelphia Wolters Kluwer; 2017

Donald Voet, Judith G Voet, Charlotte W Pratt. Fundamentals of Biochemistry. 5th Edition. New York: Wiley; 2016.

Geoffrey L Zubay, Dennis E Vance. Principles of biochemistry. Dubuque, Iowa: William C. Brown; 1995.

Jeremy M Berg, Gregory J Jr Gatto, Lubert Stryer, John L Tymoczko. Biochemistry. 9th Edition. New York: Macmillan International Higher Education: WH Freeman; 2019.

Lehninger A, Nelson D, Cox M. Lehninger principles of biochemistry. New York: Worth Publishers; 2000.

Michael A Lieberman, Rick E Ricer. Biochemistry, molecular biology, and genetics. 7th Edition. Philadelphia, Pa Wolters Kluwer; 2020.

Victor W Rodwell, David A Bender, Kathleen M Botham, Peter J Kennelly, P Anthony Weil. Harper's illustrated biochemistry. 31st edition. New York: Mcgraw-Hill Education; 2018.

CHAPTER 16

The Endocrine System

LEARNING OBJECTIVES:	**Keywords:**
• Explain the synthesis, biological significance and diseases of hormones.	Adreno-corticoids, Catecholamines, Growth hormone, Para-thyroid, Thyroid.

GH (GROWTH HORMONE)		
Fact File: GH • 19 residue polypeptide • Produced by the anterior pituitary. • Stimulates growth and metabolism in muscles, bone and cartilage cells. • Acts indirectly via growth factors (produced by the liver following its stimulation). • GH is an anabolic hormone, it stimulates protein synthesis in bone and muscle.	GH is also called somatotropin and is secreted by the anterior pituitary. The secretion of GH is mediated by two hypothalamic hormones: growth hormone-releasing hormone (HRH) and somatostatin (growth hormone inhibiting hormone). **Synthesis of GH** GH is synthesized in somatotrophs (a subclass of acidophilic cells) in the anterior pituitary gland. It is also called somatotrophin. It stimulates the production of growth factors IGF (insulin-like growth factors) in the liver. Release of GH from the pituitary is stimulated by GHRH (GH releasing hormones) and inhibited by somatostatin (SS) It is released in pulses.	
Metabolic Effects of GH		
	Decrease	*Increase*

Simmi Kharb

Carbohydrate metabolism	Glucose uptake in extrahepatic tissues Insulin sensitivity	Hepatic glucose output hepatic glycogen stores Plasma glucose
Lipid metabolism	-	Lipolysis Plasma FFA Plasma ketone bodies
Protein metabolism	Nitrogen excretion	Amino acid uptake Protein synthesis

<table>
<tr><td colspan="2">GH Disorder

GH Deficiency

It occurs both in children and adults. In children, it can be familial or may be due to tumors (cranio-pharyngioma). In adults, it is as a result of structural or functional abnormalities of pituitary.</td><td>Manifestations

In children, there is growth failure. In adults, GH deficiency occurs in case of partial or complete failure of the anterior pituitary. Symptoms include fatigue, loss of motivation, diminished feeling of well-being even social withdrawal, osteoporosis and alteration in body composition.</td></tr>
<tr><td colspan="3">GH deficiency causes a condition called dwarfism. If GH is produced in excess before bone growth stops, a person is taller than normal. This condition is called gigantism. If excess GH is released in an adult, acromegaly results. Bones cannot elongate in acromegaly, soft tissue enlargement occurs namely skin, muscle, hands, feet, nose, ears, tongue and chin.</td></tr>
<tr><td colspan="3">Acromegaly</td></tr>
<tr><td>Cause</td><td colspan="2">Pathologic GH excess

Autonomous GH excess</td></tr>
<tr><td>Pathology</td><td colspan="2">Pituitary tumor</td></tr>
</table>

	Ectopic production of GHRH or GH.
Features	Progressive enlargement of hands, feet, facial bones (including mandible, skill), coarsening of facial features. Hypertension, accelerated atherosclerosis, proximal muscle weakness, sleep apnoea, increased risk of insulin resistance/diabetes. Organomegaly is common. Excessive sweating or heat intolerance. Local effects of the tumor (headache or visual complaints) or symptoms related to other anterior pituitary hormones. Features of co-secretion of prolactin are seen in 40% of patients of acromegaly.

Thyroid Glands: Functions of Thyroid Hormone	
1. Calorigenesis:	Thyroid hormones generate heat by stimulating mitochondrial O_2 consumption and the production of ATP required for Na pump.
2. Carbohydrate and fat metabolism: (i) Carbohydrate	*Catabolic action* • Stimulate intestinal absorption of glucose • Stimulate hepatic glycogenolysis • Potentiate glycogenolytic actions of epinephrine • Stimulate insulin breakdown
(ii) Lipid:	• Lipolytic- Direct action • Indirect action via potentiating action of other hormones like: GC glucagon, GH, epinephrine. • Increase the oxidation of FFA. • Decreased plasma cholesterol by: inhibition of bile acid formation in the liver.

<table>
<tr><td>3. Growth and development:</td><td colspan="2"><ul><li>Essential for normal differentiation and maturation of fetal tissues.</li><li>In brain: myelinogenesis</li><li>Essential for GH production</li><li>Normal growth and development of the child.</li></ul></td></tr>
<tr><td>Adrenal Cortex</td><td colspan="2">Secretes glucocorticoid and mineralocorticoid hormones.</td></tr>
<tr><td colspan="2">Metabolic Effects: Glucocorticoids</td><td>Metabolic Effects: Mineralocorticoids</td></tr>
<tr><td>Carbohydrates:</td><td>stimulates GNG</td><td rowspan="6"><ul><li>Conserve Na</li><li>Regulate blood pressure</li><li>Excrete K^+, H^+, NH_4^+</li><li>Rennin angiotensin: regulate production of aldosterone.</li></ul></td></tr>
<tr><td>Lipid:</td><td>lipolysis, ketogenesis, decrease FA utilization</td></tr>
<tr><td>Protein:</td><td>increase amino acid degradation</td></tr>
<tr><td>Water and electrolyte:</td><td>Na and H_2O retention</td></tr>
<tr><td>Immune system:</td><td>inhibition</td></tr>
<tr><td>Others:</td><td>Fight, flight
HCI, pepsinogen action
Osteoporosis</td></tr>
<tr><td>Adrenal Medulla</td><td>Fact File</td><td>Hormones of Adrenal Medulla</td></tr>
<tr><td></td><td>Part of sympathetic nervous system that secretes catecholamine
Secretes epinephrine and nor epinephrine</td><td><u>Catecholamines</u></td></tr>
</table>

	Arises from neural crest Constitutes 1/10th of weight of gland Composed of chromaffin cells or pheochromocytes: large column or cells arranged in clumps or cords around blood vessels	Have a catechol nucleus with two hydroxyl side groups plus a side-chain amine Include: Epinephrine, norepinephrine, dopamine
Epinephrine	Hormone of flight and fight	
Actions	1. Vasoconstriction of arterioles (α-1 receptor) in skin, mucosa, splanchnic bed 2. Vasodilation of arterioles (β_2 receptor) in skeletal muscle 3. Bronchodilation 4. GIT smooth muscle dilation: α_2 receptor 5. Radial muscle of eye: contraction 6. Sphincter contraction: α_1 receptor, increase cardiac output.	***Metabolic Actions*** *(a) Actions mediated via:* Decreased insulin release Increased glucagon release Increased lipolysis and thermogenesis

(b) Individual actions:	*Carbohydrate Metabolism*	Increase glycogen breakdown and gluconeogenesis, raising glucose levels in brain
	Lipid Metabolism	Cause triglyceride breakdown to release FFA that serve as fuel for heart, muscle
	Physiological Effects	Increase cardiac output, BP, O_2 consumption Relax bronchus, GI Stimulate contraction of blood vessel to skin, kidney, inhibit platelet aggregation.

Norepinephrine	Major neurotransmitter of sympathetic nervous system	- Activated during fight and flight - Very little released by adrenal medulla - Cannot cross synthesize blood brain barrier, brain must synthesize them locally
Mechanism of Action	• Epinephrine act to mobilize energy in form of glucose • Inhibit amino acid release to inhibit proteolysis, sparing muscle from providing 'E' during fight and flight reaction.	
Hyperfunctioning of Adrenals	*Pheochromocytoma*	Rare tumor, more frequent in right adrenal, arise from chromaffin cells, patient presents with paroxysmal hypertension more often. *Bilateral in 10%* *10% familial* *10% malignant* *10% extramedullary*
	Paraganglioma:	Extra adrenal pheochromocytoma arising from sympathetic ganglia. Common in juxtarenal or para aortic region, more likely to be malignant Present with pain or mass.
10% of pheochromocytomas are hereditary and occur as a feature of certain familial syndromes:		*MEN (multiple endocrine neoplasia)* *MEN 2 – (multiple endocrine neoplasia)* A.D Caused by mutation in ret protooncogene
Symptoms Paroxysms of:		

Headache Sweating Forceful heartbeat Anxiety Tremor Fatigue Nausea, vomiting Abdominal, chest pain Visual disturbances		Between paroxysms: Increased sweating Cold hand and feet Weight loss Constipation ***Diagnosis*** Increase urine VMA, Metanephrine, normetanephrine. Pheochromocytoma develop in adrenals, patients tend to have hypertension, usually paroxysmal. Plasma catecholamines may be normal, metanephrine is elevated in most patients.	
MEN 2a	May develop:	Medullary thyroid carcinoma Hyperparathyroidism	Pheochromocytoma Adrenal medullary hyperplasia
MEN 2b	May develop:	Aggressive medullary thyroid carcinoma Thick corneal nerves Intestinal ganglioneuroma	Mucosal neuroma Marfanoid habitus Pheochromocytoma Adrenal medullary hyperplasia
PTH (Parathyroid Hormone)		***Parathyroid Gland***	***Actions***
Secretion		Four glands located adjacent to thyroid gland on its posterior aspect at	*PTH lowers calcium levels:*

<table>
<tr><td colspan="2">PTH is under exquisite control by serum concentration and there is negative feedback relationship of PTH with serum calcium concentration.</td><td>four poles of thyroid and exact location is variable.

Arises from 3rd and 4th brachial pouches

Composed of epithelial cells and stromal fat.</td><td>PTH regulates serum calcium levels by its effects on three target organs:

Bone

Intestine

Kidney</td></tr>
<tr><td><u>In Intestine</u></td><td colspan="3">Indirect action on intestinal absorption of calcium via vitamin D.</td></tr>
<tr><td><u>In Kidney</u></td><td colspan="3">Direct effect on tubular reabsorption of calcium, phosphate and bicarbonate (mainly in DCT).

PTH inhibits reabsorption of phosphate in proximal tubule.</td></tr>
<tr><td>Calcitonin</td><td colspan="3">32 amino acid peptides

Secreted by parafollicular C cells of thyroid derived from neuroendocrine cells from ultimobranchial body

Secretion under control of serum calcium levels

C-cells use same calcium receptor as PTH cells to sense changes in calcium concentration

C-cells increase calcitonin secretion in response to hypercalcemia</td></tr>
<tr><td>Actions</td><td colspan="3">Inhibits osteoclastic bone resorption.

Inhibits phosphate reabsorption in kidney.</td></tr>
<tr><td colspan="4">Adrenal Cortex</td></tr>
<tr><td colspan="3">Fact File

• Mesodermal in origin</td><td>Histologically composed of 3 zones:</td></tr>
</table>

<table>
<tr><td colspan="2"><ul><li>Weighs 8-10g in adult</li><li>Retroperitoneal, above or medial to upper pole of kidney.</li><li>Cortex comprises 90% and medulla about 10% of adrenal weight</li></ul></td><td>Outer: Zona glomerulosa

Middle: Zona fasciculata

Inner: Zona reticularis</td></tr>
<tr><td>Aldosterone</td><td colspan="2"></td></tr>
<tr><td>Function</td><td colspan="2">Mechanism</td></tr>
<tr><td>Conserve sodium</td><td colspan="2">Facilitate Na

Reabsorption of K^+, H^+

Secretion in DCT via renin-angiotensin</td></tr>
<tr><td>Regulate blood pressure :</td><td colspan="2">Renin released in response to falling blood volume and loss of Na^+, via renin-angiotensin system.</td></tr>
<tr><td>Aldosteronism

Cause:</td><td colspan="2">Adenoma (Conn's syndrome) idiopathic

Primary hyperplasia

Glucocorticoid suppressible

AII responsive

Carcinoma</td></tr>
<tr><td>Symptoms :</td><td colspan="2">Tiredness, muscle weakness, thirst, polyuria,</td></tr>
<tr><td>Diagnosis:</td><td colspan="2">rennin activity, plasma aldosterone, low plasma K</td></tr>
</table>

<table>
<tr><td>Biosynthesis of Hormones</td><td colspan="2">Major Hormones Secreted by Adrenal Cortex</td></tr>
<tr><td>Cortisol</td><td>Androgen</td><td>Aldosterone</td></tr>
</table>

Zona glomerulosa:	• Synthesis of aldosterone occurs here • Lacks 17α hydroxylase so can't synthesize cortisol.	
Biologic Effects of Glucocorticoids		
Target system	*Specific target*	*Physiologic function*
Metabolism	Liver	Increase expression of GNG enzyme: PEP carboxy kinase G-6-Pase F1, 6 biPase
	Adipose tissue	Elevated FFA via lipolysis (to provide fuel for GNG)
	Skeletal muscle	Proteolysis by activating ubiquitin pathway (to provide substrate (for GNG)
	Blood glucose	Maintain blood glucose during fasting Increases blood glucose level.
Calcium	Kidney	Decrease reabsorption of calcium
	Bone cartilage	Inhibit collagen synthesis, bone deposition
	GIT	Inhibit calcium, magnesium and phosphate absorption
Other endocrine	Hypothalamus pituitary	Decrease gonadotroph responsiveness to GnRH

		Stimulate GH gene expression by pituitary
	Pancreas	Inhibit insulin secretion
	Adrenal medulla	Enhance epinephrine synthesis
	Binding proteins	Decrease CBG, TBG, SHBG (all binding proteins)
Immune system	Thymus lymphocyte	Cause age-related involution of thymus
	Monocyte	Inhibit monocyte proliferation
	Granulocyte	Degranulation of neutrophils
	Inflammatory response	Inhibit inflammation by inhibiting PLA_2, inhibiting production to LT, PG. Suppression of COX-2
Skin and CT	-	Anti-proliferative for fibroblasts and keratinocytes

Disorders of Adrenal Cortex		
<u>***Adreno Cortical Insufficiency***</u> Deficient adrenal production of glucocorticoids or mineralocorticoids results in adrenocortical insufficiency.	<u>*Cause:*</u> Primary: Cause could be destruction or dysfunction of cortex *e.g.* Addison's disease Secondary: Deficient ACTH secretion.	
<u>***Addison's Disease***</u> Loss of >90% of adrenal cortex results in clinical manifestation of adrenocortical insufficiency.	<u>*Cause:*</u> Autoimmune Adrenal hemorrhage Infection Adrenoleukodystropy	Familial glucocorticoid deficiency Metastasis Drugs: Ketoconazole Metyrapone

	Congenital adrenal hyperplasia	
Cushing's Syndrome Cushing's disease is defined as specific type of Cushing's syndrome due to excessive pituitary ACTH secretion from a pituitary tumor.	Caused by chronic glucocorticoid excess, most commonly iatrogenic • Abnormality of pituitary or adrenal • Consequence of ACTH or CRH secretion by non-pituitary tumors: Ectopic ACTH, ectopic CRH	Classification ACTH- dependent ACTH – independent

ACTH- dependent type of Cushing's syndrome: Ectopic ACTH syndrome and Cushing's disease are characterized by chronic ACTH hypersecretion resulting in hyperplasia of adrenal cortex and increased secretion of cortisol, and androgen.	*ACTH – independent type of Cushing's syndrome:* May be caused by a primary adrenal neoplasia.

Clinical features: Cushing's syndrome Hyperandrogenism Hirsutism Skin: Acne due to hyperandrogenism	Obesity: Centripetal Abdominal Moon facies Buffalo hump	Gonadal: Hyperandrogenism Menstrual disorders Impotence decreased libido
Due to cortisol excess:	Musculoskeletal:	Neuropsychiatric:

Striae Bruising Plethora	Osteopenia Weakness	Euphoria Depression Psychosis

Metabolic: Renal stones Glucose intolerance Hypercalciuria	Diabetes: Hyperlipidemia Polyuria Hyperglycemia

	Addison's	***Cushing's***	***Aldosteronism***
Defect	Low GC	Raised GC	Raised aldosterone
Features	Hypoglycemia	Hyperglycemia	Alkalosis
	Loss of appetite, weight	Skin, muscle wasting	High Na
	Low BP Low Na High K Pigmentation	Hypertension Osteoporosis Negative N balance High Na Low K Buffalo hump	Low K Raised remain- angiotensin II Conn's Renal artery stenosis
Cause	Adrenal failure	i. ACTC dependent:	Primary, secondary
	Autoimmune pituitary tumor	Pituitary adenoma Ectopic	

		(loss of feedback inhibition) ACTH treatment	
	Congenital	ii. ACTH independent:	
		Adrenal tumor Adrenal carcinoma	
Ectopic ACTH is produced by:		*Adrenal function tests*	
Small cell carcinoma of lung Pancreatic islet cell tumors Pheochromocytoma	Carcinoid tumors (lung, thymus, gut, pancreas, ovary) Medullary *carcinoma* thyroid	Dexamethasone androgen suppression test ACTH test (synacthen stimulation test)	
THYROID		**Thyroid Hormone Synthesis and Secretion**	
Hormone Secreted	**Cells**	Six major steps of synthesis:	
Thyroid T_4	Follicular	1. Trapping of iodine: Active transport of I^- across basement membrane into thyroid cell. 2. Oxidation of iodide, iodination of tyrosyl residues in thyroglobulin 3. Coupling of iodotyrosine molecule in thyroglobulin to form T_3 and T_4.	
Triiodothyronine T_3	Follicular		
Calcitonin	C-cells		
Drugs Propylthiouracil, methimazole, carbimazole	**Inherited Metabolic Defects**		

inhibit TPO: Block hormones synthesis Used in management of hyperthyroiditis.	May involve any phase of hormonal biosynthesis resulting in impaired hormonal synthesis or dyshormogenesis.	4. Proteolysis of thyroglobulin, release of free iodothyronine and iodotyrosine. 5. Deiodination of iodotyrosine and conservation and reuse of liberated I. 6. Intrathyroidal deiodination of T_4 to T_3.

Iodine deficiency & Hormone Synthesis Diet low in iodine:	
Reduces intrathyroidal iodine content. Increases MIT/DIT ratio Increases T_3/T_4 ratio	Decreases secretion of T_4 Increases TSH.

Thyroid Function Tests (TFTs) Include		
• T3 (Total) • Free T3 • T4 (Total) • Free T4	• TSH • Thyroglobulin (Tg) • Anti-Thyroglobulin Antibody (Anti Tg)	• Thyroid Peroxidase Antibody (TPO) • Thyroid Antibody (Anti Tg and TPO) • TSH receptor antibody

Goitre: Any enlargement of thyroid gland is goitre.	
Simple goitre	Represents attempt to compensate for decreased thyroid hormone production, TSH is elevated
Cause	Iodide deficiency Iodide excess Defects in thyroid hormone biosynthesis

Clinical picture	Hypothyroidism
Treatment	Exogenous thyroid hormone

Hypothyroidism	Hypothyroidism in later childhood cause short stature and causes mental retardation.

Cause		***Features***	
Due to thyroid failure Disease of pituitary Disease of hypothalamus Autoimmune disease	Treatment of thyrotoxicosis with radioactive iodine/surgery Congenital Secondary hypothyroidism	BMR decreased Slow heart rate Diastolic hypertension Sluggish behavior (delayed tendon reflexes) Generalized aches	Sleepiness Constipation Sensitivity to cold Dry skin and hair (myxedema) Sallow completion

Autoimmune Disease Thyroid	
Associated with antithyroid peroxidase (antibodies): Destructive thyroid antibody causing thyroid atrophy	TSH receptor blocking antibodies: Hashimoto's disease.

Hyperthyroidism	Thyrotoxicosis due to excess production of thyroid hormone.	
Cause	Grave's disease (autoimmune thyrotoxicosis) Toxic multinodular goiter Solitary toxic adenoma	Thyroid stimulating IgG (TSI) that activate TSH®
	Drugs Secondary	Amiodarone

<table>
<tr><td></td><td></td><td>Due to TSH-secreting tumors of pituitary

Pituitary thyroid hormone resistance syndrome

Extra thyroidal hormone excess

Thyroid Carcinoma</td></tr>
<tr><td colspan="2">Clinical Features

Due to: hormone

1. Thyroid hormone excess:</td><td>Those associated with autoimmunity (seen in Grave's disease only):</td></tr>
<tr><td>Irritability

Emotional liability

Tachycardia

Atrial fibrillation

Cardiomyopathy

HT</td><td>Palmar erythema

Tremor

Diarrhea

Sweating and heat intolerance

Proximal myopathy, hyperreflexia

Clubbing of digits</td><td>Exophthalmos

Proptosis, chemosis

Lid retraction

Papilloedema

Diffuse symmetric goiter</td></tr>
</table>

PANCREAS	Made up of two functionally different organs:
Exocrine Pancreas	Digestive gland
Endocrine Pancreas	Endocrine gland, source of insulin, glucagon, somatostatin, pancreatic polypeptide.

QUESTIONS

1. Classify hormones
2. Write short note on:
   a. Mechanism of action of hormones
   b. Secondary messengers
   c. Regulation of growth hormone secretion
   d. Cushing's disease
   e. SIADH
   f. Thyroglobulin
   g. Addison's disease
   h. Pheochromocytoma
   i. Actions of parathyroid hormones
   j. Actions of vitamin D at kidney, bone and intestine
   k. Cushing syndrome
   l. thyroid hormone synthesis
   m. Synthesis of adrenal cortex hormone
   n. Biological effect of glucocorticoid
3. What are G-protein? Why are they named so?
4. What is pro-opiomelanocortin complex?
5. What is Big ACTH?
6. Differentiate between acromegaly and gigantism

7. Synthesis of adrenal cortex hormone and disorder associated with them
8. Why vitamin D is called a hormone?
9. What is the beneficiary action of estrogen on bone health?
10. What are preprohormones? Explain with example
11. Describe the mechanism of action of insulin.
12. What are different metabolic, paracrine and endocrine effects of insulin?

BIBLIOGRAPHY

Denise R Ferrier. Lippincott illustrated reviews: biochemistry. 7th Edition. Philadelphia Wolters Kluwer; 2017

Donald Voet, Judith G Voet, Charlotte W Pratt. Fundamentals of Biochemistry. 5th Edition. New York: Wiley; 2016.

Geoffrey L Zubay, Dennis E Vance. Principles of biochemistry. Dubuque, Iowa: William C. Brown; 1995.

Jeremy M Berg, Gregory J Jr Gatto, Lubert Stryer, John L Tymoczko. Biochemistry. 9th Edition. New York: Macmillan International Higher Education: WH Freeman; 2019.

Lehninger A, Nelson D, Cox M. Lehninger principles of biochemistry. New York: Worth Publishers; 2000.

Michael A Lieberman, Rick E Ricer. Biochemistry, molecular biology, and genetics. 7th Edition. Philadelphia, Pa Wolters Kluwer; 2020.

Victor W Rodwell, David A Bender, Kathleen M Botham, Peter J Kennelly, P Anthony Weil. Harper's illustrated biochemistry. 31st edition. New York: Mcgraw-Hill Education; 2018.

SUBJECT INDEX

A

B

C

D

E

F

G

H

I

K

L

P

R

S

T

U

V

W

X

Y

Z

www.ingramcontent.com/pod-product-compliance
Lightning Source LLC
LaVergne TN
LVHW070116110826
845147LV00002B/131

* 9 7 8 9 8 1 1 4 7 7 8 9 8 *